Work Measurement

Work Measurement

Harold E Dales
C Eng MIPE AMBIM MIMC
Management consultant

DISTRIBUTED IN AMERICA BY
BEEKMAN PUBLISHERS INC.
53 Park Place, New York, N.Y. 10013

Pitman Publishing

First published 1972

SIR ISAAC PITMAN AND SONS LTD
Pitman House, Parker Street, Kingsway, London WC2B 5PB
PO Box 46038, Portal Street, Nairobi, Kenya

SIR ISAAC PITMAN (AUST) PTY LTD
Pitman House, 158 Bouverie Street, Carlton, Victoria 3053, Australia

PITMAN PUBLISHING COMPANY SA LTD
PO Box 11231, Johannesburg, South Africa

PITMAN PUBLISHING CORPORATION
6 East 43rd Street, New York, NY 10017, USA

SIR ISAAC PITMAN (CANADA) LTD
495 Wellington Street West, Toronto 135, Canada

THE COPP CLARK PUBLISHING COMPANY
517 Wellington Street West, Toronto 135, Canada

Cased edition ISBN 0 273 31568 4
Paperback edition ISBN 0 273 31668 0

set in 10/11 pt. Monotype Times New Roman, printed by letterpress,
and bound in Great Britain at The Pitman Press, Bath

G2 (MAN 112/131:45)

Preface

The aim of this book is to present the main techniques of work measurement and method study and show how they can be marshalled into action by following a procedure which has economic cost reduction as its dominant theme.

It is hoped that it will serve to enlighten managers, work study practitioners and students alike as to the tremendous possibilities of work study when it is used aggressively to attack the waste factor—and to reaffirm its status as one of the most powerful tools at the disposal of managements today.

I am indebted to a fellow management consultant, Colonel W. Grigor, who urged me to write this book, and to my wife, without whose constant help and encouragement it would never have been completed.

I would like to acknowledge that the description of the Work-Factor system in Chapter 13 was based on information taken from the *Industrial Engineering Handbook* by H. B. Maynard, published by the McGraw-Hill Book Company.

My thanks are also due to the Methods-Time Measurement Association Limited for permission to reproduce the *MTM*-1, *MTM*-2 and *MTM*-3 Data Cards in Chapter 13.

London
May 1972

H. E. D.

Contents

Preface   v

1 The work study department   **1**
Work study techniques   1
Broad analysis techniques   2
Fine analysis techniques   3
Micro-analysis techniques   3
Methods of using work study   4
Convincing management   4
Convincing supervision   5
Convincing labour   5
Announcing the introduction of work study   6
Establishing a work study department   6
The size of the work study department   9
The position of the department in the management structure   10
Relations with other departments   11

2 Fundamentals of work measurement   **14**
Demonstration of the rating method   15
Rating scales   18
The concept of standard effort   18
Division of work into elements   20
Extension of observed times   21
Checks on rating   28
The work unit   30

3 Time study   **31**
Equipment   31

                                                                      vii

Contents

Timing methods   32
Time study observation procedure   32
Time study summarizing procedure   39
Examples   43

4 The selection of basic element times   **47**
Procedure for the selection of basic element times   48

5 Relaxation allowances   **60**
The computation of relaxation allowances   60
Arrangement of data   61
Physical strains   61
Mental strains   63
Working environment   65
Minimum relaxation allowance   67
Extra allowance for females   67
Relaxation allowance computation sheet   67
Typical examples of RA taken from practice   67
Outdoor work   69
Relaxation allowance used in example   69
Tea breaks   69

6 The compilation of basic synthetic data   **71**
Procedure   71
Note on example   84

7 Checking the accuracy of measured work   **85**
The computation of complete standard times per operation   85
Checking standard times   88

8 Broad work measurement   **95**
Work measurement using an ordinary wrist watch or clock   96
Work measurement by direct observations without the use
   of a time piece   98
Work measurement from examination of work records   98
Analytical estimating   99
Activity sampling   101
Accuracy and number of observations for activity sampling   108
Activity sampling procedure   110
Rated activity sampling procedure   116
Analytical observation   117

Predetermined motion-time systems 122
Practical applications 122

9 The measurement of team work **125**
Unoccupied time and process allowance 125
Multiple activity charts 125
Procedure for the measurement of team work 128
Example 1—Reduction of unoccupied time on a team of two
   workers 131
Example 2—Reduction of unoccupied time on a "flow-line"
   team 131
Interference losses 136

10 The measurement of process-controlled work **137**
Definitions 137
Illustrative example 139
Exploration of existing working methods 139
Process time study 141
Numerical example of process time study 144
Determination and checking of net standard times 145
Determination of unoccupied time on a single machine or
   process 148
Reduction of unoccupied time 149
Determination of unoccupied time on multiple machine work 151
Calculation of machine interference allowances 157
Determination of ancillary allowance 163
Analysis of operation under theoretically optimum conditions 164
Determination of final standard times 171

11 Final presentation of data **173**
Work specifications 173
Storage and referencing of standards data 176
Use of special computation forms 177
Accuracy of standard times 179
Register of time values 182
The effect of learning and interruption 183
Estimating the time necessary for work measurement 184

12 Method study **186**
Method study procedures 187

Contents

Preparation techniques 187
Collection techniques 188
Presentation techniques 197
Process chart symbols 197
Preparation of data for the construction of process charts 197
Constructing process charts 200
Preparation of data for constructing flow diagrams and string
    diagrams 205
Constructing flow diagrams 205
String diagrams 205
Multiple activity charts 205
Simultaneous action (simo) chart symbols 207
Preparation of data for the construction of simo charts 208
Constructing simo charts 209
Cyclegraphs and chronocyclegraphs 209
Document analyses 209
Material audits 212
Process analysis 212
Analysis techniques 212
Development phase 217
Preparation for installation 217
Consolidation techniques 218

13 Predetermined motion-time systems **219**
Work Factor 220
Methods-time measurement (MTM) 224
Second generation systems 227
MTM-2 235
MTM-3 237
Specialized data systems 238
Tape data analysis 239

14 Incentives **240**
Introduction 240
Incentives for direct labour 241
Incentives for indirect labour 256
Incentives for supervision 257
Learning allowances 258
The choice of incentive system 259
Launching an incentive 260
Revision of incentive conditions 261

15 Labour cost control   **262**
Direct work   262
Measured work   263
Uncontrolled work   263
Ancillary work   263
Excess work   263
Waiting time   263
Allowances   263
Operator performance   263
Department performance   264
Overall performance   264
Performance and rating   265
Attendance time   265
Standard costs   265
Actual costs   266
Excess costs   266
Make-up   267
Overtime premium   269
Control sheets   269

16 Overhead and material costs   **270**
Labour cost   271
Material cost   271
Overhead cost   271
Classification of overhead cost   272
Allocation of overhead cost   272
Marginal cost   274
Costing methods   274
Other terms used   274
Determination of product cost   275
Overhead recovery accounts   276
Fixed overhead recovery   277

17 Cost reduction reports   **278**
Reference periods   278
Direct labour reference   278
Reference periods inclusive of indirect labour   280
Reference period inclusive of overheads   281
Machine utilization reference periods   281
Reference period used for method study applications   282

Contents

References for material utilization   282
Surveys   282
Survey reports   285
Savings calculations   287
Cost reduction reports   290

18 Summary of procedures   **293**
Part 1—Introduction   293
Part 2—The procedure in outline   294
Part 3—The procedure in detail   298
Part 4—Referencing procedure   315
Part 5—Adaptations of the procedure to suit different appli-
     cations and conditions of work   322

19 Other techniques   **333**
Production planning and control   333
Project network analysis   334
Process planning   336
Variety reduction   336
Value analysis   336

Bibliography   **338**

Glossary of terms   **340**

Index   **347**

1

The work study department

Work study is simply the study of work. It is no new idea. The realization that efficient work performance is sound economy must surely date to the beginnings of civilization. It is probably this knowledge which has led to the development of highly skilled crafts and trades. The nature of the techniques used to study and develop these activities is largely unknown. But it must have been based on one simple concept—the application of thought to action. What is now known as work study is no different in principle—its character has merely become more sophisticated because of the increased complexity of working methods. Due to the fierceness of competition and the need for continual review of internal economy, it has now become an accepted and firmly implanted management aid.

Work study techniques

Work study is the analysis of work into small parts followed by rearrangement of the work pattern and conditions to give the same effective result at less cost. It examines both the method and duration of the work, and can be said to encompass two separate techniques—method study and work measurement. No work study investigation can be entirely thorough, however, unless it employs both of these, as method, time and cost are inseparably interdependent.

Because both the conditions under which work is studied as well as the character of the work itself will vary with every investigation, different sub-techniques have been evolved. The choice of which to use for any situations will be dictated by the cost of applying them relative to the financial return. They can be classified by degree as follows—

1 Broad analysis techniques.
2 Fine analysis techniques.
3 Micro-analysis techniques.

Each one of these is now examined in more detail.

Broad analysis techniques

These are generally suitable for investigating work of a non-repetitive nature, such as maintenance, building, tool-making and office work; or stores and factory layout, transportation problems, etc. They can also be applied to situations where conditions are unusually chaotic, this being a first step in an overall improvement project which will concentrate on finer detail at a later stage.

Method study techniques would include—

(*a*) Process planning, which is the division of work into specialized technical processes, and the design of tools to reduce the element of skill.

(*b*) The construction of flow diagrams to represent work operations and facilitate subsequent analysis. These are known as outline process charts, two and three dimensional flow process charts and string diagrams, according to their form.

(*c*) Variety reduction, which is the reduction of unnecessary variety of product of work operation by simplification.

(*d*) Multiple activity charts, which are the representation of work operations in the form of a bar chart.

(*e*) Network analysis, which is a means of diagrammatically representing complex operations where some of the individual tasks can be performed simultaneously.

Work measurement techniques would include—

(*a*) Analytical estimating, which is the determination of time values by skilled appraisal.

(*b*) Activity sampling, which is the determination of work values and delay intervals by statistical sampling.

(*c*) Timing by wrist watch or wall clock, which is the determination of time values of relatively long duration by direct observation and rating.

(*d*) Second generation pre-determined motion-time data, by which use is made of known time values of the fundamental movements of the human body reduced to broad terms to build up times.

(*e*) Evaluation from synthetic data. This is a method of obtaining operation times from special purpose data developed by using some or all of the above methods to suit particular types of work.

Outline process charts, two and three dimensional flow process charts and string diagrams are further discussed in Chapter 12; process planning and variety reduction in Chapter 19; multiple activity charts in Chapter 9; and broad work measurement techniques in Chapter 8.

Fine analysis techniques

These are the most widely used techniques and are suitable for the investigation of work of a repetitive or semi-repetitive routine. They are also useful for consideration of improvements to workshop and workplace layouts; tool design; and as an aid to the selection of machinery and plant.

Method study techniques would include—

(*a*) Two-handed process charts, which are charts constructed from conventional symbols to show the activities of both hands in performing an operation.

(*b*) Multiple activity charts.

Work measurement techniques would include—

(*a*) Time study by stop watch, which is the evaluation of work duration by timing and rating intervals of fairly short duration—usually from 0·1 to 0·5 minutes.

(*b*) Predetermined motion-time data which has an accuracy of more than ±0·5 per cent. Amongst the more commonly known systems are MTM-1 and Work Factor.

(*c*) Evaluation from synthetic data. This is the determination of operation times from special purpose data developed by using time study, predetermined motion-time data, or a combination of both of these.

Two-handed process charts are further discussed in Chapter 12 and predetermined motion-time data systems in Chapter 13. Evaluation from synthetic data is explained in Chapter 11.

Micro-analysis techniques

These are generally used to study very short cycle repetitive operations, where small reductions in process time may realize relatively large cost saving.

The method study technique used is known as motion study. This employs charts called simultaneous motion cycle (simo) charts to record detailed movements of the fingers and hands in performing work operations. The technique was invented by the American Consultants, Frank and Lilian Gilbreth, who used a standard symbol called a "therblig" (which is an anagram of Gilbreth) to represent the fundamental elements of movement, and it is further discussed in Chapter 12.

Work measurement techniques include analysis by ciné-film. These have been used mainly to develop the predetermined motion-time systems explained in Chapter 13.

Methods of using work study

Applying work study to a problem can sometimes be a difficult and complex procedure. In order for it to be successful, therefore, it is essential to assess the benefits beforehand; and from then on to choose the most appropriate techniques to secure them at an economical cost. It must never be assumed that any one of these existing techniques is superior to all the others. The real skill is in selecting the best to suit the conditions. New techniques are always emerging, which according to their originators supersede most of those in current use but again these claims should never be accepted. Any new technique should be tried, fairly judged and then placed in its proper perspective—as yet another tool for the work study analyst.

With these considerations in mind, a detailed summary of procedures which give guidance in selecting the most suitable techniques is given in Chapter 18.

Although direct cost reduction should always result from the successful application of work study, there are also other advantages. Incentive bonus systems based on work measurement are always the more successful and if properly administered can lead to better industrial relations. They also allow the introduction of labour cost control. This often reveals other weaknesses in an organization and shows the advantages of installing such techniques as production control, planned preventive maintenance, improved inspection services, etc.

Convincing management

Management inexperienced in the use of work study may be sceptical of its effectiveness. It is therefore essential that all initial application should take place in an area where sufficiently impressive results can be obtained in a relatively short time. It is important to know how to select these, and also be able to compile and present reports to management which clearly set out the method of approach, the time required, and the expected financial returns based on a proposal which can be readily understood and easily demonstrated.

4

Management will usually accept the principle of work study when it has produced positive results. Because of this, the initial approach should be made by an experienced and competent practitioner who is not only able to present a case logically and in a convincing manner, but is also able to substantiate his claims by a subsequent application within the specified time.

Convincing supervision

Supervision and medium level management are perhaps the most difficult group of people to convince. This is often due to resentment. They may feel that the introduction of the technique is a criticism of their past performance. If it has been decided to introduce work study, it should be explained that this is not with the intention of criticizing, but merely to bring greater security and prosperity to the organization as a whole—and to them as part of it. They should be persuaded to give the work study practitioner all possible help to allow the results of the work to be brought to fruition quickly.

Supervision easily become disgruntled if they are overlooked in the general improvements that come from the application of work study. Incentive bonus schemes may mean increased earnings to operators, and management will benefit from the lowering in operating costs, whereas supervision have to suffer a complete upheaval in their department, often with no reward whatsoever. This situation should be anticipated, and a share of the savings made should be set aside for them in return for their co-operation. The investment will be well worth while.

During installation of the new methods a sympathetic attitude should be maintained, for a good deal of resentment can be overcome by painstaking and honest explanations and the maintenance of respect for first-hand technical knowledge. Very often valuable suggestions will be made by supervision and due credit should always be given to the originators of these by higher management. Far better end results will be achieved by this means than if the work study analyst attempts to claim all the credit himself.

Convincing labour

The introduction of the technique should be preceded by open discussion with the trade union representatives. Operations will be subjected to a considerable amount of close study and investigation and this will lead to strained industrial relationships and suspicion unless the workers know quite clearly what is to take place. Management should give firm guarantees that no actual changes will take place until these have been discussed and agreed with the union representatives. If it is the intention to introduce bonus incentives,

5

this should be mentioned, but the extent of the reward to be offered should not be disclosed until work study is complete, and agreement has been drawn up beforehand.

In the same way, further extensions to work study operations should be discussed, so that labour will be ready to accept changes more readily.

Announcing the introduction of work study

The intention to introduce work study techniques should be announced simultaneously to supervision and labour and if not at the same time, most certainly on the same day. This is most important, as suspicion and resentment will grow if the news leaks out through the wrong channels. It is convenient to appoint a small committee of representatives of supervisors and workmen, chaired by a top executive or director at which a full explanation is given. If it is the first introduction of the technique to an organization, it should be followed by posting a brief notice on the factory notice boards.

Establishing a work study department

The prime function of a work study department is to submit proposals and implement schemes to increase productivity and reduce waste. It must, therefore, be capable of—

(a) Recommending and implementing improvements to factory layout, working conditions, working methods, tooling, product design, office organization and clerical methods.

(b) Producing measured work values.

(c) Initiating and maintaining incentive bonus schemes.

(d) Setting standards for labour cost and originating labour cost control documentation which shows deviations from these standards in the form of excess costs.

The basic needs of the department are very simple, particularly during the initial stages. They merely consist of the normal office equipment—desks, chairs, filing cabinets and stationery. Dependent on the scope of the activities, however, some thought should be given to specialized equipment and specially designed stationery. This last consideration is important, as it will not only reduce clerical effort, but provide the basis for an efficient filing system. The referencing procedure is explained in Chapter 18, part 3. The equipment needed for carrying out special techniques such as time study, predetermined motion-time study, and method study is explained in the appropriate chapters headed with these descriptions. For further guidance, most of these are listed below—

(*a*) *Time study* (reference Chapter 3)
  (i) instantaneous fly-back stop-watches;
  (ii) study boards.
 (iii) work study observation sheets (Fig 9);
 (iv) work study observation and record sheets (Fig. 10);
  (v) work study top sheet (Fig 13);
 (vi) study register sheet (Fig 11).

(*b*) *Predetermined motion-time systems* (reference Chapter 13)
  (i) analysis sheets (Fig 61);
  (ii) portable tape recorders if tape data analysis is to be carried out;
 (iii) equipment as for time study in addition.

(*c*) *Method study*
  (i) process chart sheet (Fig 46);
  (ii) ciné-camera with counter and related equipment if motion study is to be carried out (see Chapter 12);
 (iii) still camera and chronocyclegraph equipment if these are to be taken (see Chapter 12);
 (iv) equipment as for time study in addition.

(*d*) *Data summary for work measurement*
  (i) study summary—constant element sheets (Fig 17);
  (ii) study summary and register of variable elements sheets (Fig 18);
 (iii) graph sheets—large and small types (Fig 19);
 (iv) register of constant elements sheets (Fig 21);
  (v) collection of elements sheets (Fig 22);
 (vi) data sheets—large and small types (Fig 25);
 (vii) relaxation allowance computation sheets (Fig 20);
(viii) set up register sheets (Fig 15).

(*e*) *Data summary for method study*
  (i) motion analysis sheets (Fig 51);
  (ii) graph sheets and data sheets as given under (*d*) (iii) and (vi) above.

All the forms should be on the same sized paper (A4). This will greatly facilitate filing.

Absolute neatness and accuracy must always be insisted upon throughout the whole department, both in the execution of the clerical work and the subsequent filing.

The head of the work study department is usually described as a chief work study analyst or officer, who, in addition to being well versed in method study and work measurement, should have a good appreciative knowledge of cost accounting, with particular reference to the significance of overhead expenses and budgeting. Although not essential, it is better if he understands the rudiments of the technical processes which are carried on within the plant. A good general background of training in mechanical engineering principles is also a decided asset, particularly in an engineering works, since many of the investigations will invariably criticize the process planning and general tooling methods in current use.

The qualifications of such a man will vary widely according to the size of the department he controls, and the nature and scope of the work he is authorized to carry out. But one quality is paramount—diplomacy. For work study means change, and change implies criticism—which will cause resistance and resentment if it is not handled carefully. His manner of approach should therefore at all times be honest, tactful and persuasive, and he should be well able to deal with personnel at all levels and implement the necessary changes without causing undue friction.

Although his function is purely advisory and carries no executive authority outside his own department, he should be capable of dealing with management at the highest levels and constantly submitting purposeful and authoritative reports on proposals for cost reduction and subsequent savings claims.

Other personnel in the department may include work study analysts (or officers); work study observers; and clerical staff. Work study observers are usually junior members whose duties are mainly the collection of data from the activity centres under investigation. They should be well trained in time study and method study collection techniques, and be competent arithmeticians. Neatness is also essential, and tidiness and orderliness in clerical filing should be insisted upon.

The work study analysts collate, analyse and devise alternative proposals under the guidance of the head of the department, and should work to a large extent under their own initiative. Together with the chief work study analyst, they should be trained in mathematics at least to "O" level or its equivalent—preferably higher, if problems involving complex statistical analysis are necessary. It is a mistake to divide the department sharply into the two main techniques of work measurement and method study, for to be a competent team, the staff should be alive to both these aspects of work study.

Because the work study department initiates bonus incentive schemes and labour cost control, there will always be a certain amount of clerical work necessary in conjunction with these. Functionally, however, these belong in the accounts department, and they should be transferred to that department as soon as the preliminary stages of installation of new techniques are complete.

Clerical staff can relieve work study officers of many simple calculations such as extension of basic times, general filing duties, report typing, as well as calculations for bonus incentive and labour cost control which occur in the early stages of development of new schemes.

The size of the work study department

The number of personnel of the work study department is largely dependent on the size of the industry, the complexity of the work involved, and the extent to which the department is used by management. Assuming plants of equivalent labour strength, the factors tending to increase the numbers of study staff will be the amount of non-repetitive work, the amount of change in product design and manufacture, the scope for the use of synthetic work measurement, and the degree of good organization present before the introduction of the technique.

Typical strengths for an industry employing 1,000 personnel are given below as a guide—

Repetitive work

Chief work study engineer	1
Work study engineers and observers	2–3
Clerical staff	2–3

Non-repetitive work

Chief work study engineer	1
Work study engineers and observers	3–5
Clerical staff	3–4

If, however, the department is used to investigate such aspects as documentation (organization and methods), production control installations, network analysis, value analysis, operational research, etc., more staff will be required.

The criterion for determining the size and composition of the department is economy. To be effective, the department should justify its existence by being able to recover its own cost in terms of financial savings within a reasonably short time. This should be judged not only on material results, but on the ability to maintain these in the future. Once the department is installed, it must be used to maintain the improved conditions and actively encouraged to function during the stages of future planning, in order that potential waste can be attacked before it occurs.

9

The work study department

The position of the department in the management structure
The work study department should be integrated into the management team as soon as possible, with the scope of its responsibilities carefully defined. Although the position will largely depend on the size of the organization and the importance placed on the function, it is usually convenient to incorporate it within the works manager's administration. A typical structure of a small engineering works of approximately 1,000 strong is shown in Fig 1.

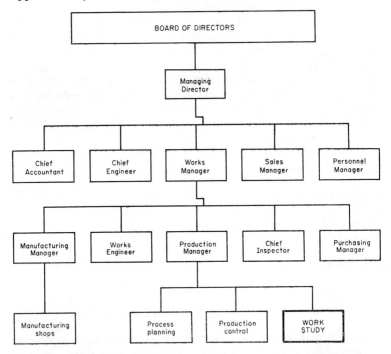

Fig 1 Typical management structure of an engineering works of approximately 1,000 strong, showing the position of the work study department

Here, the managing director directly controls five executives; a chief engineer (concerned with product design and development); a chief accountant; a sales manager; a personnel manager and a works manager. The works management structure consists of a manufacturing manager or chief works superintendent who controls the manufacturing shops; a chief inspector; a works engineer (responsible for maintenance of plant and buildings); a purchasing manager; and a production manager. The production manager (who may be alternatively described as a chief industrial engineer) would be responsible for a chief process planning engineer, a production controller and chief work study officer.

This enables the production manager to co-ordinate the related services of *process planning*, which is the analysis of work into separate technical operations and the sequential listing of these on an *operation layout; work study*, which will subject all operations and working methods to close scrutiny, minimize the work content, and publish measured work values; and *production control*, which will use operation layouts with work values added to them as a basis for machine and operator loading. It also enables the works manager to use work study as a service to assist in factory layout, work place layout, internal transport facilities, etc.

Work study should never be stifled in its scope by being buried too far down the management structure. With this in mind many larger organizations set up a department reporting directly to the board of directors under a chief industrial engineer or head of management services, specifically to assist in major planning for future production, control of systems and documentation (organization and methods, vetting of capital purchase for new machinery and plant, etc.). Under this scheme, work study is offered as a service to the rest of the organization, some of the personnel being "attached" to the works manager's administration if this is more convenient for co-ordination of the related functions described above.

A structure of this type is shown in Fig 2.

Relations with other departments

Work study is a service to the rest of the organization—but it should be controlled and administered by line management. This is why the work study department should test any proposed schemes for economic viability and present them to management on this basis before any action is taken. Also, after the detailed investigations are complete, line management action will be again necessary to authorize the final recommendations and insist that they are carried out. The work study department is powerless without this, as it has no executive authority outside its own department for the very obvious reason that it would not be able to function properly if it had. It would become inundated with routine enquiries which would stifle constructive thought and render it almost sterile.

Work study may create personnel problems—particularly where there are bonus incentive payments. Since the personnel department is responsible for the preservation of good industrial relations, work study officers should keep the personnel manager informed on all plans which affect labour. Clear explanations of any proposed

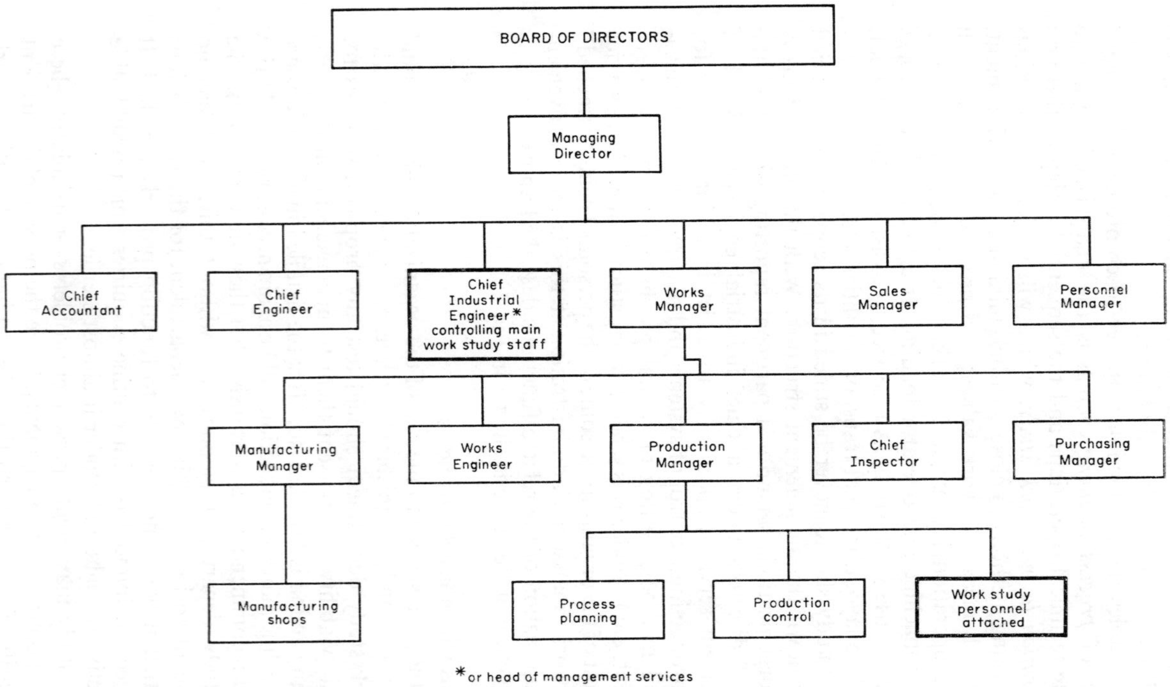

Fig 2 Management structure of a large engineering works. Here the work study services are grouped together under a chief industrial engineer (or head of management services) and some of the work study staff are attached to the production manager's staff for convenience of working

* or head of management services

changes should be given to operators in the presence of the shop steward and the personnel manager. If a dispute arises over a measured work value because of low bonus earnings, every effort should be made to settle the grievance—even to the extent of offering to take a long production study to prove the fairness of the work values, working closely with the personnel department throughout the proceedings. An attitude of fair-mindedness is essential for the continued success of work study, and if a mistake has been made, it should be admitted freely and rectified immediately. By this means, it will become more quickly accepted as an established technique within the organization.

2

Fundamentals of work measurement

One of the main difficulties of measuring work is the variability in length of time when there are variations in effort, even if precisely the same method is used each time. It is quite impossible for a worker to maintain a constant effort over any but a very short length of time, but if a standard effort could be defined such that could be maintained on average throughout a working day, work could be expressed in units of time with respect to this standard.

One method of calculating time values in this way is by straight averaging of a considerable number of observations. A study could be made on several operators who could be relied on to work at a steady, even rate over a period; and the arithmetic mean of the observations calculated. This, however, would clearly be costly and difficult in practice. But if observers could be trained to recognize a standard effort, far fewer times would need to be recorded. Furthermore, if they could also be trained to assess deviations from standard to close limits, even fewer observations would be necessary, as the observed times could be adjusted to give the duration of each at standard effort. Under these conditions, specially selected operators would no longer be necessary.

Time study is based on this theory, variations in effort being expressed in points values according to some conventional scale. This assessment is known as *rating*.

Various rating scales exist, but the one now favoured by the British Standards Institution uses 100 as the standard effort. Under this system, assessments are carried out between 50 and 150 degrees of effort in steps of 10 (optionally 5 when close to the 100 or standard effort). The scale is no longer practical beyond these limits. Accuracies of $\pm 2\frac{1}{2}$ per cent should normally be possible by experienced observers when a number of observations have been averaged, although there are exceptions when certain craft work is involved.

After taking observations by this means, it is necessary to convert the recorded values to basic times (i.e. the actual times which would have been taken if the work had been performed at the standard rate). This process is known as *extension* and is calculated from the formula.

$$\text{Basic time} = \frac{\text{observed time} \times \text{observed rating}}{\text{standard rating}}$$

For example, if an operation is timed at 20 seconds duration and the British Standard scale of rating is used to rate the operator's effort at 90 (the standard rating being 100), the basic time would be—

$$\frac{20 \times 90}{100} = 18 \text{ seconds}$$

or, in other words, the operator would have completed the operation in 18 seconds if he had worked at standard effort.

There are various different methods of obtaining the average of a series of basic times taken on the same work operations. These are explained later in the chapter.

Demonstration of the rating method

Some people inexperienced in the rating method are sceptical of its practical value. The real way to convince them is to apply time values which have been obtained by its use to operators under incentive conditions. If the group of operators is sufficiently large (say 10 or more), average actual times should be approximately equal to the standard times when allowance has been made for relaxation. But it is not always practical to carry out such a demonstration, and in these cases, a simple experiment can be used instead.

A pack of ordinary playing cards is taken and a number of people asked to sort them as quickly as possible into four piles of thirteen, as if dealing for whist or bridge. The duration of each deal is recorded with a stop watch to the nearest whole second, mis-deals being discounted. At the same time, a considerably smaller number of deals are timed and rated by an independent observer who is skilled in this technique and the average basic time of his set of

15

readings then compared to the average of the observed times of the previous, larger, set.

Providing that a sufficiently large number of deals are made, the rating method can be said to be effective if the two results are within a small percentage of each other.

Duration (seconds)	Number of times each duration occurs			
	1st sample	2nd sample	3rd sample	Totals
16	1	—	—	1
17	—	1	2	3
18	—	3	4	7
19	6	8	7	21
20	9	13	12	34
21	14	19	17	50
22	20	25	20	65
23	16	21	16	53
24	11	6	8	25
25	8	3	6	17
26	6	—	5	11
27	4	—	2	6
28	3	1	—	4
29	—	—	1	1
30	1	—	—	1
31	—	—	—	—
32	1	—	—	1
Totals	100	100	100	300
Arithmetic mean	22·9 seconds	21·6 seconds	22·0 seconds	22·2 seconds

NOTE: The arithmetic mean was obtained in each case by multiplying each duration by the number of times it occurred, adding the results together and dividing by the total number of observations.

Fig 3 Chart of observations taken on a number of people dealing out four hands of bridge

A summary of the recorded times of ten people dealing a pack of cards ten times, each on three separate occasions, is shown summarized in Fig 3. The arithmetic mean of each of the three sets of 100 observations is seen to vary, being respectively 22·9, 21·6 and 22·0 seconds, and the mean of the whole 300 readings, 22·2 seconds.

A sample 25 observations was made simultaneously by a skilled time study observer who rated each deal for speed and effort. These are shown summarized in Fig 4, recorded times being extended by multiplying them by the rating and dividing by 100. The arithmetic mean of the basic times was 22·3 seconds.

Actual duration (seconds)	Speed and effort rating	Basic time (seconds)	Number of observations	Total basic (seconds)
19	120	23	1	23
20	110	22	2	44
21	110	23	2	46
21	100	21	1	21
22	100	22	7	154
23	100	23	5	115
23	90	21	1	21
24	90	22	3	66
25	90	23	2	46
26	80	21	1	21
		Totals	25	557
		Average basic time (seconds)		22·3

NOTES: The "100 at standard" rating scale was used. Basic times were calculated by multiplying by the speed and effort rating and dividing by 100 in each case.
Total basic times were obtained by multiplying them by corresponding numbers of observations. The arithmetic mean of the basic times was calculated by dividing the total of all basic times by the total number of observations.

Fig 4 Chart of observations taken by a skilled time study observer on dealing playing cards during the experiment charted on Fig 3

Results are summarized below, times being expressed in seconds and minutes for further clarity—

	Straight average method	*Rating method*
Number of observations	300	25
Average time (seconds)	22·2	22·3
Average time (minutes)	0·370	0·372
Per cent error (rating method)	—	0·5%

From this it can be seen that accuracy to within $\frac{1}{2}$ per cent was obtained by using the rating method as against the straight average method, but only one-twelfth of the number of observations was necessary.

Rating scales
Fig 5 compares the different rating scales in common use. As has been mentioned before, the 100 scale is now favoured by the British Standards Institution (BSI).

Efforts can briefly be described as—

50	poor
60	very slow
70	slow
80	fair
90	good
100	standard
110	standard +
120	fast
130	very fast
140	exceptionally fast
150	limit of human effort

The concept of standard effort
In order to rate successfully, the concept of standard or optimum effort must be clearly understood. It can be described as that effort which is the highest that one would reasonably expect an average operator to exert and maintain throughout a working day. Speed, however, is influenced both by the amount of physical effort and the care and attention necessary, assuming comparative skill. The exertion of standard effort when doing heavy physical work or delicate adjustments involving high concentration may appear to be slow and ponderous when compared to simple, well organized assembly operations, but they should all nevertheless be rated

18

RATING SYSTEM	100	1·00*	60/80	100/133
Standard rating	100	1	60	100
Rating carried out to nearest	10 units (5 close to 100)	0·1 units	5 units	10 units
Units of work developed	Standard times	Standard times	Allowed times	Allowed times
To obtain basic time multiply observed time by observed rating and divide by—	100	1	60	100

Comparative Scales	Practical Rating Limits ↑ ↓	Absolute Rating Limits ↑ ↓				
			—150—	—1·5—	—120—	—200—
			—125—	—1·25—	—100—	—166—
			—100—	—1·00—	—80—	—133—
			—75—	—0·75—	—60—	—100—
			—50—	—0·50—	—40—	—66—

Fig 5 Comparison between different rating scales in common use

similarly, as they require the same comparative effort on the part of the operators.

Some work is restricted by the process itself, such as spraying paint with a spray gun. Providing that the apparatus is working efficiently and the paint is being directed properly and not excessively, there is little more that the operator can do to influence the speed of the operation. Under these conditions he should be rated at standard or the time classified as "process time".

In those conditions where an operator is obliged to wait until a

* This system of rating is commonly used in Europe.

machine has completed its cycle—such as may happen on a lathe when the automatic traverse is operating—the time interval is classified as ineffective time. Any work elements performed during the cycle are timed and rated separately, some indication being necessary to record the point at which the machine cycle starts and finishes.

Other work may widely vary in time even if the effort is constant. This commonly occurs where a fair amount of skill and judgement is required; where there are unavoidable fluctuations in material or process, such as occur on polishing, soldering, applying adhesive, etc.; or in craft industries such as pottery, instrument manufacture, etc. In these cases the operations should be carefully observed and rated according to the amount of effort applied, providing it is satisfactorily established that the fluctuations in time are unavoidable and that there is not a deliberate attempt being made to deceive the observer.

Although constant vigilance must be maintained to ensure that work is being performed in a proper manner and with reasonable effort, the analyst should always retain an attitude of fair-mindedness when rating, otherwise values which are "tight" or "loose" will result and one of the prime objects of work measurement will be defeated.

Rating assessments are made with reference to the standard, i.e. whether the effort has been greater or less and to what extent. They should be sufficiently sensitive to the inevitable variations of effort, or the rating will be too "flat", i.e. have too great a tendency to be within a narrow range either side of the optimum. This is a common fault with beginners (see "checks on rating" at the end of this chapter).

Division of work into elements

No human being can maintain an even pace of working for very long. There will be inevitable fluctuations of effort even when the steadiest workers are performing highly repetitive work.

Because of this, rating can only be carried out successfully over relatively short intervals of time. The practical limit is about three-quarters of a minute or 80 centiminutes.

On the other hand, if very short intervals of time are taken this can lead to inaccuracies in timing. For example, an error of half a centiminute in two centiminutes represents an inaccuracy of between 25 per cent and $33\frac{1}{3}$ per cent, whereas the same error in a 20 centiminute interval is only $2\frac{1}{2}$ per cent.

If work is observed closely, it will be noticed that most of it will be quite naturally divisible into intervals which range from 10 to 80 centiminutes long. When carrying out time study, therefore, it is usual to break down the work in this way, timing and rating each interval separately.

These intervals are known as *elements*.

Extension of observed times

It has been mentioned earlier that observed times are converted to basic times by a process known as *extension*. This uses the following formula—

$$\text{Basic time} = \frac{\text{observed time} \times \text{observed rating}}{\text{standard rating}}$$

For the 60/80 system this is—

$$\text{Basic time} = \frac{\text{observed time} \times \text{observed rating}}{60}$$

And for the 100/133 and the 100 British Standard system—

$$\text{Basic time} = \frac{\text{observed time} \times \text{observed rating}}{100}$$

The selection of average basic time from several observations made of the same element can be carried out by several different methods. Some of these are—

(*a*) Extending observed times individually and calculating the arithmetic mean of the results.
(*b*) Extending observed times in groups and calculating the arithmetic mean of the results.
(*c*) Selecting the modal average.
(*d*) Using the reciprate graph.

These are now considered in more detail.

(*a*) *Extending observed times individually and calculating the arithmetic mean of the results*

Individual basic times can be calculated in several ways. One method is to use a slide rule set at each rating in turn, move the cursor to observed times on one scale and read off the corresponding basic times on the other.

Another method is to construct a table which gives basic times for various ratings. This is used by first finding the observed time under the standard rating column and reading off the basic time under the appropriate rating column. A specimen part of such a table is shown in Fig 6.

21

2

Ratings

50	60	70	80	90	95	**100**	105	110	120	130	140	150
3	3	4	4	5	5	**5**	5	6	6	7	7	8
3	4	4	5	5	6	**6**	6	7	7	8	8	9
4	4	5	6	6	7	**7**	7	8	8	9	10	11
4	5	6	6	7	8	**8**	8	9	10	10	11	12
5	5	6	7	8	9	**9**	9	10	11	12	13	14
5	6	7	8	9	10	**10**	11	11	12	13	14	15
6	7	8	9	10	10	**11**	12	12	13	14	15	17
6	7	8	10	11	11	**12**	13	13	14	16	17	18
7	8	9	10	12	12	**13**	14	14	16	17	18	20
7	8	10	11	13	13	**14**	15	15	17	18	20	21
8	9	11	12	14	14	**15**	16	17	18	20	21	23
8	10	11	13	14	15	**16**	17	18	19	21	22	24
9	10	12	14	15	16	**17**	18	19	20	22	24	26
9	11	13	14	16	17	**18**	19	20	22	23	25	27
10	11	13	15	17	18	**19**	20	21	23	25	27	29
10	12	14	16	18	19	**20**	21	22	24	26	28	30
11	13	15	17	19	20	**21**	22	23	25	27	29	32
11	13	15	18	20	21	**22**	23	24	26	29	31	33
12	14	16	18	21	22	**23**	24	25	28	30	32	34
12	14	17	19	22	23	**24**	25	26	29	31	34	36
13	15	18	20	23	24	**25**	26	28	30	33	35	38

Fig 6 Specimen rating table

A third and probably the quickest method is to calculate basic time mentally using the following guide—

Rating	Guide
80	Multiply by 8 and move decimal point
90	Multiply by 9 and move decimal point
110	Add a tenth
120	Add a fifth

Extension is only necessary to the nearest whole number. To prove this, consider the example shown on next page.

The averages are 25·13 and 25·10 respectively. The difference is negligible and any slight gain in accuracy does not warrant the extra effort involved. Work measurement is a practical technique, not an exact science.

The process of averaging can be speeded up if a basic time is selected at random, this successively subtracted from each value,

Observed time (Centiminutes)	Rating	Basic time to one decimal	Basic time to nearest whole number
28	90	25·2	25
26	95	24·7	25
25	100	25·0	25
29	90	26·1	26
25	100	25·0	25
26	100	26·0	26
28	90	25·2	25
30	80	24·0	24
31	80	24·8	25
23	110	25·3	25
	Totals	251·3	251·0

the arithmetic mean taken of the result and this added to the random selection (or subtracted, if it is negative).

To illustrate this, use the example above and select a value of 25 centiminutes. Now subtract this successively from each value, average the result, and add it to the original 25. The result is as follows—

$$25 + \frac{0 + 0 + 0 + 1 + 0 + 1 - 1 + 0 + 0}{10}$$

$$= \quad 25 + \frac{1}{10}$$

$$= \quad 25\cdot1 \text{ as before}$$

With practice, this can become a very quick mental calculation.

(b) *Extending observed times in groups and calculating the arithmetic mean of the results*

This gives the same results as the above method but it tends to be quicker, especially where a fair amount of observations have been taken.

Observed times given the same rating are added together and extended in groups. The sum of these is then divided by the total frequency to give the arithmetic mean.

Applying this to the above example gives results as follows—
(extension has been carried out to the nearest whole number)

Ratings	80	90	95	100	110
Observed times	30	28	26	25	23
	31	28		25	
		29		26	
Totals	61	85	26	76	23
Basic times	49	76	25	76	25

The total basic time = 251 centiminutes
The number of
  observations   = 10
Average basic time = 25·1 centiminutes
which is the same result as before.

(c) Selecting the modal average

This consists of selecting the most frequently occurring basic time
and can be an even quicker method. It is particularly suitable for
highly repetitive work, but it should only be used by experienced
observers as it needs a certain amount of skilled judgement.

Consider the following example—

Observed times (Centiminutes)	Rating	Basic times
25	100	25
25	100	25
25	100	25
24	110	26
25	100	25
24	100	24
26	90	23
32	80	26
25	100	25
25	100	25

These could be tabulated as follows—

Basic time	Frequency
23	1
24	1
25	6
26	2

The most frequently occurring time or modal average is 25 centi-minutes. This would be the value selected by this method.

The arithmetic mean is 24·94 centiminutes which is an error of less than 0·25 per cent.

A still quicker method is to select the most frequently occurring observed time and rating—

Observed times and ratings	Frequency
24 @ 100	1
24 @ 110	1
25 @ 100	6
26 @ 90	1
32 @ 80	1

The most frequently occurring time and rating is 25 centiminutes at 100 rating, which gives 25 basic centiminutes as before.

In practice there is usually no need to tabulate the results. An experienced observer can scan observed times and ratings and make the selection direct from the observation sheet.

(d) Using the reciprate graph

The formula for extension from observed time to basic time can be expressed as—

$$b = tr$$

where b = the basic time

t = the observed time

r = the rating

But basic time b is constant. If therefore, the reciprocal of the rating is used, the formula can become—

$$b = t \div \frac{1}{r}$$

$$\text{or} \quad t = \left(\frac{1}{r}\right) b$$

25

This means that if observed times are plotted against ratings arranged on a reciprocal base, the points should lie on a straight line passing through the origin.

The method is shown on Fig 7. A convenient scale is drawn on the base line and observed times plotted against their respective ratings to give sufficient points on the graph to enable a straight line to be drawn through them and the origin. The selected basic time is then read off where line A cuts the 100 line. Line B on the graph has been

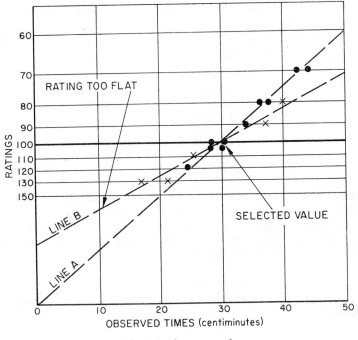

Fig 7 Reciprate graph

drawn to illustrate how checks on rating can be carried out and is explained later in the chapter. This is its main use. It is too slow for normal time study work.

Some element times may be exceptionally high when compared to the bulk of those recorded. There may be several reasons for this. Unavoidable fluctuations in work content would be one of them, such as occurs frequently in polishing operations. Another reason could be that the operation is not being performed consistently due to poor tooling. In extreme cases it could also be a deliberate attempt being made to deceive the observers.

The problem is whether to include extreme values in the average and a useful test for this is to determine whether the full range of values with the high one included is twice that when it is excluded. For example, consider the following basic times—

22, 26, 24, 23, 21, 32, 26.

The value 32 appears to be out of line.
To test this fully, first find the full range.

Full range = highest value—lowest value
= 32—21
= 11

Next, find the range with the high value excluded.

= second highest value—lowest value
= 26—21
= 5

Since 11 is more than twice 5, the high value 32 appears to be suspect.

If no legitimate reason can be seen for this anomaly, it should be excluded from the values when calculating the arithmetic mean. On the other hand, if there is a necessity for such increases in work content due to the nature of the operation, a different action should be taken.

Consider an element described as "polish on buffing spindle" referring to polishing a metal casting. This gave a reasonably consistent time over 51 observations of 47·4 basic centiminutes, but there were 4 separate readings which in themselves averaged 79·6 basic centiminutes. Further investigation revealed that this was due to unavoidable fluctuations in the surface of the material. Two elemental times were therefore recorded, i.e.—

Polishing on buffing spindle
= $47 \cdot 4^{55}$

Extra polishing due to inconsistency in material quality
= $79 \cdot 6 - 47 \cdot 4 = 32 \cdot 2^4$

The index figure inserted against these values indicates the frequency of observation. It is a convention used throughout this book, and should indicate the *actual* number that took place during the study period, despite the fact that less may have been used for averaging.

For example, if a time of 39·5 basic centiminutes was obtained by averaging 29 observations, but the element actually occurred 31

times during study, the selected average should be recorded as
39·5[31] and not 39·5[29].

Checks on rating
Accuracy in rating is fundamentally important to work measurement. It is therefore essential that observers are competent in this technique.

A method of checking rating consistency and accuracy is to study a man walking.

Mark a course out, say 100 feet long over flat, level ground free from obstructions. Ask a man to walk over this at different speeds. He should have normal walking shoes and should already be in motion when crossing the starting and finishing lines.

Since a normal man between the ages of 20 and 60 of average height and in good health should walk at the rate of 4 miles per hour on flat, level ground when exerting an effort of 100 BSI, this can be used as a standard. For a course of 100 ft—

Rating	*Time*
80	35·5 centiminutes
90	31·6 centiminutes
100	28·4 centiminutes
110	25·8 centiminutes
120 etc.	23·7 centiminutes

Every observer should rate 10 times without using a watch. Actual times should be recorded for each pass and the ratings calculated from the standard of 28·4 centiminutes at 100 BSI.

When comparison is made the *average* error should not be more than $\pm 2\frac{1}{2}$ per cent.

Other methods which can be used are the examples of dealing cards given earlier in this chapter, or rating films which can be bought or hired.

Another method is to plot observed times against rating on a reciprate graph (Fig 7). A line is drawn through the plotted points ignoring the origin (line B). If it slopes away from the normal line (line A) then the rating is too "flat"—a common fault with beginners.

Work measurement is a practical tool, not an exact science. Individual elements cannot be assessed to high accuracy; this is only obtained by taking the average of several basic times.

Rating limits are stated below. Attempts at greater accuracy are both impractical and unnecessary.

60/80 system

From 40 to 60 and 100 to 120, rate to the nearest 10 points; between 60 and 100, to the nearest 5 points.

100/133 system

Rate to the nearest 10 points over the full range from 70 to 200.

100 (British Standard) system

From 50 to 90 and 110 to 150, rate to nearest 10 points; between 90 and 100, rate (optionally) to the nearest 5 points.

The 100 British Standard system is used throughout this book to maintain consistency.

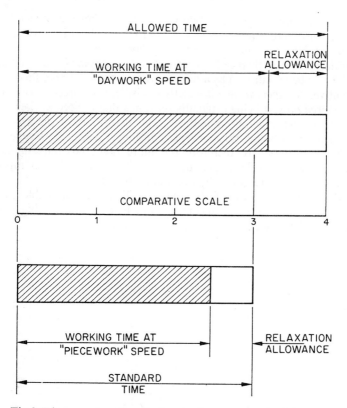

Fig 8 Diagram of work units showing how one standard time unit is equivalent to three-quarters of an allowed time unit. (Although this ratio still remains true, the concept of allowed time based on a "daywork speed" is no longer accepted as being valid.)

The work unit

The performance of work causes fatigue and the cure is rest, the greater the strain the work or working conditions impose, the longer the rest needed. Time is also needed for personal hygiene. To arrive at operation times that can be maintained throughout a working day therefore, allowance must be made for these. Addition of such allowances as a percentage of the basic or optimum time develops what are known as standard units of work, operation times being expressed in these units. A work unit is diagrammatically illustrated in Fig 8 and can be described as—

A fraction of work plus a fraction of rest, the proportions of each varying according to strain but the total always aggregating unity.

Work units are expressed in two ways—*allowed times* and *standard times*. An *allowed* time is based on that effort which is supposedly maintainable throughout a working day by an average worker working without financial incentive, and a *standard time* is that which can be maintained by a piece worker or worker under financial incentive. A *standard time* is exactly equivalent to three-quarters of an *allowed time*, it being originally reckoned that a piece worker will maintain rates which are one-third faster than those of the average day-worker, although this concept is no longer accepted as valid.

Standard times are therefore the more favoured units in use. *Allowed times* are developed using the 60/80 and 100/133 rating methods, which are gradually becoming obsolete.

Work units, because of their invariability, are invaluable as a basis for bonus incentive schemes, work loading and costing, and are the ultimate units for work measurement.

3

Time study

Equipment

The equipment needed for time study is extremely simple. It consists of a stop watch, an independent time piece (such as a wrist watch or wall clock), a study board and a series of forms on which to record the information. The stop watch should be of the instantaneous fly-back type fitted with a centrally pivoted hand which sweeps the main dial. A small dial and pointer mounted off-centre records the number of revolutions of this central hand.

There are two controls, a spring loaded winder and a starting lever at the side. Moving this lever towards the stem will start the hands in motion. Depression of the winder will instantly return them to zero, and if the lever is in the start position, the watch will re-start when it is released. If the winder is depressed in the stop position, the hands will return to, and remain locked at, zero.

Stop watches are obtainable graduated in seconds, decimal minutes or decimal hours. Seconds graduations are more easily explained and will arouse less suspicion amongst operators, but those graduated in decimal minutes or decimal hours cause less subsequent arithmetic. To be consistent, the decimal minute type is used for examples throughout this book, the graduations on the dial being in one-hundredths of a minute or *centiminutes*.

The wrist watch or clock used as an independent time piece must be fitted with a seconds hand, preferably mounted centrally.

The study board is merely a piece of shaped wood, hardboard or other suitable material on which observation forms can be clipped. There should also be a means of supporting the stop watch on the board using a special spring clip or similar device.

Observation forms will be either of two types, according to the nature of the work being timed. For non-repetitive work use *work study observation sheets*, and for repetitive work, *work study observation and record sheets*. Examples of how these two forms are used in this way are shown in Figs 9 and 10 respectively.

Timing methods

Two systems of recording are used in time study, the flyback and cumulative methods.

When timing by the cumulative method, the stop watch is allowed to run on continuously, readings being written down to mark the starting and final termination points of each time interval.

With the fly-back method, the winder of the stop watch is rapidly depressed and released at starting and final termination point of each interval, allowance being made for a slight loss in time taken for the hand to be returned to zero and be re-started again by taking fractional readings and recording to the next higher value. For example, an interval of exactly 30 centiminutes would be recorded as precisely 30 centiminutes, whereas an interval of 30 centiminutes and a fraction would be written down as 31 centiminutes and so on.

The cumulative method is used for broad measurement by direct timing (see Chapter 8). It is not generally recommended for time study as it then requires a good deal of arithmetic to obtain the individual time intervals.

Fly-back timing must always be checked with the independent time piece as explained in the procedure which follows.

Time study observation procedure

Before any time studies are taken, the investigation should be frankly discussed with the foreman or overseer on equal terms and his advice sought on the selection of the best operators for preliminary study. It should be explained that they must be properly skilled in their work, the most suitable being those who are by nature co-operative.

Accurate time standards are difficult to obtain by studying operators who continually work below 80 British Standard rating. Although fluctuations of effort are both natural and unavoidable, the best subjects to study are undoubtedly those who work steadily at rates close to standard (i.e. BSI 100).

The procedure should then follow that as set out below, except that slightly less emphasis may be placed on the preliminary approach to the operators when there has been more general acceptance of the techniques within the area under review—

WORK STUDY OBSERVATION SHEET

Product __4" diameter brass case__   Study No. __51__

Operation __Drill ½" diameter hole 1" deep.__   Sheet No. __3__ of __5__

Operator __W. Bryant__   Date __19.7.71__

Element	Ref				Element	Ref			
Start at 11.22 a.m.					Procure new brush	(G)	81	100	81
I.C.T.		35	-		Lubricate drill	D	12	80	10
Load to jig	A	13		13	Brush away chips	E	21	100	21
Drill ½ dia hole	B	36		23		A	12		13
Unload from jig	C	48	100	12		B	33		23
	A	12		13		C	44	110	12
	B	32		22		A	14		14
	C	43	110	12		B	37		23
	A	14		14		C	49	100	12
	B	37		23		A	14		14
	C	49	100	12		B	36		22
Lubricate drill	D	10	90	9		C	48	100	12
	A	14		13	I.T.		15	-	
	B	39		23	Talk to foreman	(H)	90	100	90
	C	52	90	12	Lubricate drill	D	9	90	8
	A	13		13		A	12		13
	B	35		22		B	32		22
	C	47	100	12		C	43	110	12
Brush away chips	E	25	90	23		A	12		13
	A	12		13		B	33		23
	B	33		23		C	44	110	12
	C	44	110	12	Brush away chips	E	19	100	19
I.T.		12	-			A	13		13
Lubricate drill	D	9	100	9		B	35		22
	A	14		14		C	47	100	12
	B	37		23	Lubricate drill	D	8	100	8
	C	50	100	13		A	14		13
	A	13		14		B	40		23
	B	32		21		C	53	90	12
	C	43	110	12		A	12		13
	A	14		13		B	32		22
	B	39		23		C	43	110	12
	C	52	90	12		A	14		14
Procure cutting oil	(F)	62	90	56		B	36		22

Fig 9 Type of observation sheet used for studying work where there are numbers of regular elements or where occasional elements occur at random. Elements were given letter references during study except those ringed. The study commenced at 11.22 a.m. and the initial check time was 35 centiminutes. Fly-back timing was used except for elements *A*, *B* and *C* which were timed by the continuous method

WORK STUDY OBSERVATION AND RECORD SHEET

Product __10" dia brass case__   Study No. __202__

Operation __Drill ¼" dia hole x 1" deep.__   Sheet No. __2__ of __2__

Operators   J. Lawson   Date 28·7·70

Start at 10.32 a.m.							A			B		C		
I.C.T.	35	-	-				25	100	25	42	100	20	110	23
							26	95	25	41	100	22	95	21
							28	90	25	40	105	21	100	21
							24	105	25	42	100	21	100	21
							26	100	26	42	100	22	100	22
							25	100	25	44	95	25	80	20
							25	100	25	42	100	21	100	21
							27	90	24	42	100	21	100	21
							23	110	25	40	105	22	95	21
							27	95	26	42	100	22	100	22
							22	110	24	42	100	23	90	21
							25	100	25	42	100	21	100	21
							24	100	24	46	90	19	110	21
							25	100	25	41	100	21	100	21
							24	105	25	42	100	22	100	22
F.C.T.	88	-	-											
Finish at 10.43 a.m.														

	A	B	C
Check	total	select	total
time			
total =	374^{15} =	$42 @ 100$	319^{15} =
123·0	$24·9^{15}$ =	$42·0^{15}$	$21·3^{15}$

Fig 10 Combined observation and record sheet used to study a highly repetitive operation. Letter references were inserted at the head of the columns before study. Elements *A* and *C* were extended individually and the arithmetic mean taken, but element *B* was selected from a modal average of 42 at 100

(*a*) Discuss the investigation with the operators to be studied, the workers' representative being in attendance if requested.

(*b*) Observe the operation for a period and analyse it into separate work elements.

(*c*) Synchronize the stop watch with the independent time piece, record the time of day and the initial check time.

(*d*) Time and rate every work element and record all other happenings.

(*e*) Synchronize the stop watches with the independent time piece, record the time of day and the finishing check time at the conclusion of the study.

(*f*) Have the quality of the work checked and record all relevant details of the operation and the work-place.

(*g*) Thank the operator for his co-operation.

Each of these steps is now explained in detail.

(*a*) *Discuss the investigation with the operators to be studied, the workers' representative being in attendance if requested*

Operators selected for study should be introduced to the work study observer by the foreman. An explanation as to what is to take place should then follow. Initially, it is wise to invite the shop steward or other representative to be present. This may not be necessary during subsequent investigations when the operators have become used to the technique.

No effort should be spared to explain in simple terms that the intention at this stage is merely to measure the work, and no further action will be taken until there has been full discussion and agreement between management and the workers' representative on any further moves. It should also be pointed out that time studies are at liberty to be examined at any time, should this be so wished.

(*b*) *Observe the operation for a period and analyse it into separate work elements*

Most work consists of a series of distinctly separate elements some of which occur regularly every operation cycle and others which are more infrequent.

For example, using a vertical drilling machine to drill a hole in a piece of metal which is clamped in a jig to hold it steady, may contain elements which occur regularly every cycle as follows—

1 Procure the piece of metal from a supply container and clamp in jig.
2 Drill one $\frac{1}{4}$ inch hole, 1 inch deep.
3 Unclamp and remove piece from jig and lay aside to disposal container.

Others may occur regularly, but not every cycle, as—

1 Lubricate the drill with cutting medium.
2 Remove metal chippings from drilling machine table.
3 Procure empty container.
4 Dispose of full container.

Some may happen very infrequently, as—

1 Procure supply of cutting medium.
2 Procure replacement brush.
3 Receive instructions from supervision, etc.

Elements in the first group are called *repetitive elements*, since they occur regularly each operation cycle; those in the second group, *regularly occasional elements* as they occur regularly but not every cycle; whilst those in the third group which do not appear according to any cyclic pattern, are termed *irregular occasional elements*.

Work measurement should not commence until the operation has been observed over several cycles in order to decide how best it can be analysed, the principles to be borne in mind being—

(i) Elements must have a clearly defined *break point* which is audible, if possible, such as the sound of an article being tossed into a container, or the release of a drill spindle after drilling a hole.

(ii) Variable length elements are separated from those of constant length.

(iii) Elements involving light work are segregated from those which are of a strenuous nature.

Some work may consist of a few repetitive elements with very few, if any, occasional elements. These can be described in some fair detail with indication of the break points using a *work study observation and record sheet* for the purpose. Each element is given a letter reference, these being placed at the head of the columns before the actual timing is started (see Fig 10).

Other work, however, involves a greater number of elements per cycle well interspersed with occasionals. These operations will need to be watched for a longer period until the rhythm has been fully absorbed and understood. The elements themselves will have to be briefly described as they occur during the timing process using a *work study observation sheet* (Fig 9).

(*c*) *Synchronize the stop watch with the independent time piece, record the time of day and initial check time*
The observer being equipped with a reliable watch or clock fitted

with a seconds hand, a stop watch and a supply of *observation forms* clipped to a study board, should place himself in such a position to be able to see the operator's movements clearly. The stop watch should be held in line with his eye and the centre of activity. Preferably, the observer should not stand behind the operator but should be clearly visibly by him.

He should then wait until the seconds hand on the independent time piece reaches the zero or "12 o'clock" position and at that precise moment start the stop watch by operating the slide. The time of day is then recorded at the top of the *observation form*.

At the very point at which the timing is commenced (which will normally be at the beginning of a work cycle or working day), the stop watch reading is noted and the hand returned to zero. It is then re-started by rapidly depressing the winder and instantly releasing it, the reading being recorded on the *observation sheet* as *ICT* (initial check time). An example of this entry is seen in Fig 9.

(*d*) *Time and rate every work element and record all other happenings*
Every incident which occurs during a study must be recorded by description and time interval, periods where the operator is not working being accounted for as *IT* (ineffective time) or *RA* (relaxation allowance or rest period).

The fly-back timing method should be used, times being recorded to the nearest whole centiminute as previously explained.

Ideally, element lengths should lie between 10 and 50 centiminutes (6 and 30 seconds). Where they are appreciably longer than this, they should be artificially sub-divided into intervals of not more than 50 centiminutes, each being separately rated. Shorter elements can be recorded by using the continuous timing method over a small group. For instance, if elements of 5, 15 and 7 centiminutes occur in sequence, these would be written down as respectively 5, 20 and 27 centiminutes, the whole three being rated together and the watch returned to zero at the end of the interval. Individual times can then be obtained later by subtraction.

It is immaterial whether the time is entered first and the rating afterwards on the *observation form*. Some practitioners advise the rating to be written down before the time is entered so that the observer will not be influenced by the watch reading. This is regarded as being quite unnecessary, as any work study observer who would be swayed in this way is either insufficiently trained or unsuitable for the occupation.

Frequently the observer must measure operations on which two or more operators are working, such as occur in the work of electricians, bricklayers or fitters and mates, etc. In these cases, each man is studied separately and any elements which are performed together, such as lifting operations, etc., described as this on the *observation form*.

(*e*) *Synchronize the stop watch with the independent time piece, recording the time of day and the finishing check time at the conclusion of the study*

The study will normally be closed at the end of an operation cycle or working period. When the last element has been entered, the stop watch is left running until the seconds hand on the independent time piece has reached the zero or 12 o'clock position. At this very instant, the stop watch is halted by operating the slide, and the reading recorded on the *observation sheet* as *FCT* (finishing check time), together with the time of day as shown on the independent time piece (see Fig 10).

(*f*) *Have the quality of the work checked and record all relevant details of the operation and the workplace*

It is essential that all work measured gives results which are within acceptable quality standards, and arrangements should be made to have it checked by a competent authority. Quality which is too high is as unacceptable as that which is sub-standard, although an occasional falling-off is sometimes unavoidable due to fluctuations in quality of material, working conditions or human limitations. If such incidents are rare, the extra work necessary should be accepted as part of the operation as a contingency; but if they are frequent, they should be further investigated with management and supervision until they have been reduced to a practical working minimum.

It is often wise to have the quality certified by an inspector's signature, as it may be subsequently needed to support claims that measured work values apply only to work of acceptable quality.

Quantities produced should also be properly substantiated, particularly if a long study has been taken which may later be used to confirm work values. In these cases it may be advisable to weigh check the production during the study.

Other details should also be recorded at this stage such as—

(i) Description of the product.
(ii) Description of the plant used, if any.
(iii) Description of any special tools used.
(iv) Operating speeds of machines, if these are used.
(v) Product part numbers, operators' names and works numbers, etc.

(vi) The number of operating cycles (i.e. strokes of a press, vat loads in a dipping process, etc.).

(vii) Description of the operation.

(g) Thank the operator for his co-operation

The simple courtesy of thanking the operator for his co-operation should never be forgotten, as it will go far in establishing the good relations which are so essential for the success of any work study application.

Time study summarizing procedure

Time studies should be summarized as soon as possible after they have been taken whilst the information is fresh in the mind of the observer. The procedure for so doing can be reduced to a routine as follows—

(*a*) Enter brief details to the *study register*, and allocate the time study a reference number.

(*b*) Allocate a common reference letter to each element on the time study.

(*c*) Summarize and select basic element times.

(*d*) Summarize the check, ineffective and relaxation times.

(*e*) Calculate the elapsed and net effective times.

(*f*) Estimate the average rating.

(*g*) Enter all relevant details to the *work study top sheet* and staple all observation sheets together in sequence.

This is described in detail below.

(a) Enter brief details to the study register and allocate the time study a reference number

Brief details should be entered to a study register (Fig 11), from which the study is given a reference number. This is then entered to the head of each observation sheet at this stage, and on each *summary sheet* and *top sheet* subsequently.

(b) Allocate a common reference letter to each element

Most of the regularly occurring elements will have been allocated a reference letter during the progress of the study, particularly if highly repetitive work was being observed, and a *work study observation and record sheet* used for recording the data.

During this summarizing stage, however, the study should be carefully examined and *all* elements marked. Examples of this procedure are shown in Fig 9, the reference letters added after the study was taken being shown ringed.

STUDY REGISTER		
Serial No.	Description	Date

Fig 11 Study register

(c) Summarize and select basic element times

Basic element times are calculated or selected from the observed times by extension. The various methods of doing this were explained in Chapter 2.

(d) Summarize the check, ineffective and relaxation times

The actual values of the check, ineffective and relaxation times are extracted from the observation sheets and totalled using *work study observation and record sheets*, (Fig 12) for the purpose.

Where long production studies are taken lasting a half-day or more, the ineffective time (or time when the operator is not working), should be segregated into relaxation time and other ineffective time. These should also be separately totalled on the summary sheets.

(e) Calculate the elapsed and net effective times

The elapsed time is the difference between the starting and finishing times of the study. For example, if the study commenced at 8.32 am and finished at 12.02 pm, then the elapsed time E—

$$= 12.02 - 8.32$$
$$= 3 \text{ hours } 30 \text{ minutes}$$
$$\text{or } 210 \text{ minutes}$$

The net effect time is the elapsed time minus the sum of the check, relaxation and ineffective times. Using a numerical example to explain this—

If c = the check time (say, for example, 123 centiminutes)
n = the relaxation time (say, 739 centiminutes)
i = the ineffective time (say, 2,448 centiminutes)

WORK STUDY OBSERVATION AND RECORD SHEET

Product __4" diameter brass case.__   Study No. __51__

Operation __Drill ¾" dia hole by 1" deep.__   Sheet No. __2__ of __5__

Operators   W. Bryant.   Date 19·7·71.

A	B	C		D		E		F	G
Load to Jig	Drill ¾" dia hole	Unload from jig		Lubricate drill		Brush away chips		Procure oil	Procure brush.
$13^{32} = 416^{32}$	$21' = 21'$	$12^{38} =$	456^{38}	$8^6 =$	48^6	$19^2 =$	38^2	$56'$	$81'$
$14^8 = 112^8$	$22^{19} = 418^{19}$	$13^2 =$	26^2	$9^5 =$	45^5	$20^2 =$	40^2		
	$23^{20} = 460^{20}$			$10' =$	$10'$	$21^3 =$	63^3		
						$23' =$	$23'$		
Total = 530^{40}	Total = 899^{40}	Total = 482^{40}		Total = 103^{12}		Total = 164^8			
Av = 13.3^{40}	Av = 22.5^{40}	Av = 12.1^{40}		Av = 8.6		Av = 20.5^8		Av = $56.0'$	Av = $81.0'$

H	I	K	Check times		Ineffective times	
Talk to foreman	Aside full pallet.	Get empty pallet				
$90'$	$17'$	$15'$	ICT	35	12	
			FCT	12	15	
					14	
Av = $90.0'$	Av.$17.0'$	Av = $15.0'$	Total = 47.0		Total = 41.0	

Fig 12 A work study observation and record sheet being used to summarize a time study and select the basic times by arithmetic mean. After extension, identical basic times were entered together with the frequency with which they occurred during study shown as index figures. These were then extended to sub-totals, the sub-totals and indices being separately added. Division of total basic times by frequency (index) gives the selected basic times

Time study

then the net effective time e would be—

$$e = E - \frac{(c + m + i)}{100}$$

in the *example*—

$$e = 210 - \frac{(123 + 739 + 2448)}{100}$$

$$= 176 \cdot 9 \text{ minutes}$$

(f) Estimate the average rating

The average rating R is calculated by dividing the total basic centiminutes B by the net effective time e or—

$$R = \frac{B}{e}$$

using an example—

Where $B = 18{,}603$ centiminutes

and $e = 176 \cdot 9$ minutes

$$R = \frac{18603}{176 \cdot 9}$$

$$= 105 \text{ BSI}$$

This method is too lengthy for general use, however, and an experienced analyst should be able to estimate this value within $\pm 2\frac{1}{2}\%$ by carefully scanning the ratings recorded on the *observation sheets* if good, steady operators are selected for study. Average ratings should be expressed to the nearest whole number.

(g) Enter all relevant details to the work study top sheet and staple all observation sheets together in sequence

The *work study top sheet* (Fig 13), should now be completed with the following details—

(i) Department
(ii) Section
(iii) Product
(iv) Operation
(v) Number of cycles. (This is the number of cycles of a machine or process, where this is relevant.)
(vi) Output. (This is the number of pieces, lb., gallons or other relevant units produced during the study period.)

42

(vii) Plant details. (The type of machine, etc., together with operating speeds, etc.)
(viii) Study number
(ix) Date
(x) Taken by (enter here the observer's initials).
(xi) Operator's name
(xii) From (time the study was started). To (time the study was finished).
(xiii) Estimated average rating
(xiv) The elapsed time
(xv) The check time
(xvi) The ineffective time
(xvii) The net effective time

Items (xiv) to (xvii) should be entered in minutes to the nearest first decimal point only. Further accuracy is quite unnecessary.

Every element observed during study is also entered to the *top sheet* at this stage. Each should be carefully referenced by letter to the *study observation* and *summary sheets* in such a way that any element can immediately be traced to these sheets. Details of break points for the most frequently occurring elements should also be entered.

Basic times should normally be entered in centiminutes to the first decimal place only. Frequencies should record the exact number of times the element was performed during the study—even if all these were not actually written down on the observation sheets. For example, if a particular element was timed and rated 35 times during a study, but two extra elements were performed which were not rated due to faults in the process or the mode of operation, the full frequency of 37 would be recorded on the top sheet.

It is most essential that the top sheet is completed extremely neatly and legibly so that it can readily be understood by personnel other than the observer who took the study.

After completion of all the entries, the *data sheets* should be arranged in sequence with the observation sheets underneath, the summary sheets next, and finally the top sheet at the head. Each sheet should then be numbered in sequence and stapled together.

Examples

Examples of a completed *observation sheet*, *summary sheet*, and *top sheet* are given in Figs 9, 12 and 13 respectively. Observations entered to these sheets refer to the study of a simple drilling operation. It consists of placing metal castings in a drill jig, drilling one hole $\frac{1}{4}$ in. dia. by 1 in. deep in each, and laying them aside to a disposal container. Fig 14 is an example of a *data sheet*.

WORK STUDY TOP SHEET

Department _Machine Shop_	Study No. _51_
Section _Drilling_	Date _19.7.70_
Product _4" dia brass case_	Taken by _K.J.W_
Operation _Drill ¼" dia_	Operator _W. Bryant_
hole by 1" deep.	From _11.22_ to _11.47 a.m._
	Estimated Av. Rating _103_
No. of cycles _40_ Output _40_	Elapsed time _25.0 mins_
Plant details _Sensitive drill_	Check time _50 centiminutes_
Remarks _Total error in_	Ineffective time _40 centiminutes_
timing checked as +0.3%	Net effective time _24.1 mins_
	or 0.42 hours.

Element reference	Element	Basic time centimins.	Frequency	
A	Procure case from pallet & load to jig	13.3	40	
B	Drill ¼" dia hole	22.5	40	
C	Unload case from jig to pallet	12.1	40	
D	Lubricate drill with cutting oil	8.6	12	
E	Clear metal chips from drill table	20.5	8	
F	Procure cutting oil from stores	56.0	1	
G	Procure new brush from stores	81.0	1	
H	Discuss work with supervision	90.0	1	
J	Aside full pallet to storage area	17.0	1	
K	Move empty pallet to loading area	15.0	1	

Fig 13 Work study top sheet. This summarizes all the information collected during study. It should be completed extremely neatly and legibly. The extra column can be used to indicate element "break-points" to ensure consistency in element lengths

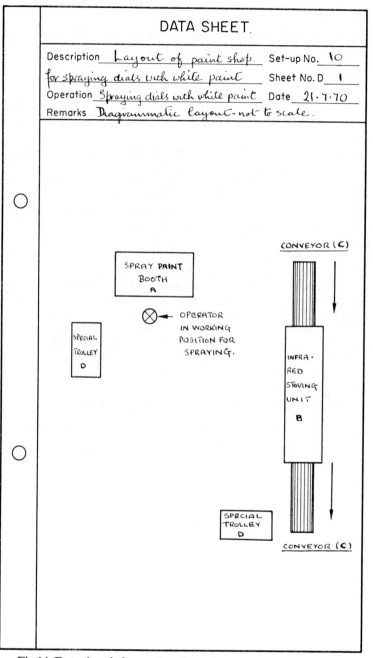

DATA SHEET.

Description Layout of paint shop. Set-up No. 10
for spraying dials with white paint Sheet No. D 1
Operation Spraying dials with white paint Date 21·7·70
Remarks Diagrammatic layout - not to scale.

CONVEYOR (C)

SPRAY PAINT
BOOTH
A

⊗ ← OPERATOR
IN WORKING
POSITION FOR
SPRAYING.

SPECIAL
TROLLEY
D

INFRA-
RED
STOVING
UNIT
B

SPECIAL
TROLLEY
D

CONVEYOR (C)

Fig 14 Data sheet being used to show diagrammatic layout of an
operation for spraying dials with white paint. The form is a
general purpose sheet having a variety of other uses in addition

Chapters 4 to 7 describe procedures for the determination of standard times. To assist the explanation of these, another example is referred to. This is "spraying dials with paint", and is described as follows.

Tinplate discs ranging from 2 in. to 12 in. diameter are withdrawn from stores in batches of 500 to 2,000 in size. These are taken into a paint shop where there are a number of waterwash spray paint booths.

The operation consists of picking up these discs one at a time, carefully wiping them clean, and spraying them with a white stove enamel. The sprayed dials are then carefully laid face upwards to a wire mesh tray. When the tray is full of dials it is placed on a moving conveyor which moves it slowly under an infra-red heating device, which bakes the enamel hard. Trays of finished stoved dials are then loaded to a trolley which is specially constructed to take several of these. Full trolleys are then wheeled to the next department to be printed with markings to make them into clock dials.

Occasionally, trolleys become empty in the printing department and are wheeled into a position convenient for loading with fresh trays of dials, empty trays being taken from them as required.

The operator is also required to obtain supplies of enamel from the paint store, filter it through muslin when required, and replenish the spray gun with enamel when it becomes empty. From time to time also, the spray gun becomes blocked and has to be washed with a paint solvent to clean it.

The operation is shown diagrammatically in Fig 14, in which A is the spray booth; B the infra-red stoving unit fitted with a slowly moving conveyor C, arranged to take the special tray; and D is a trolley which has been designed to take 10 of the trays.

4

The selection of basic element times

A single job of unique nature is usually made up of a number of constant work elements combined together in various ways. Standard times for such work can be obtained quite simply. Basic element times are first selected from time studies and relaxation allowances added to them. Then the frequency with which they occur in the main operation is determined, each element being multiplied by these frequencies and the results added together.

Work of such isolated character, however, rarely occurs in practice. More often than not there is other work which is basically similar. Analysis of a range of jobs in a particular area reveals that they consist of a surprisingly small number of different types of element which occur with different frequencies, some of which vary according to different characteristics of the work and the product, and others which remain constant.

If, therefore, standard times for these elements can be established, and the variations in time and frequency connected with some easily recognizable feature of the product or process, standard times for a whole range of different types of work can be built up without the need for further time study.

This has great advantages over the method of obtaining standard times by individually timing each operation. Apart from being more economical on time study, it ensures that all operation times are in true proportion to each other. Moreover, it often means that standard times can be computed before new work is released, which is useful for estimating as well as ensuring that the proportion of measured work is kept at high level.

The procedures outlined in this and succeeding chapters are therefore designed to compile *synthetic data,* i.e. elemental standard times and frequencies arranged in such a way that they can be used to build up standard times for complete operations by synthesis. The method of arriving at standard times for operations of a unique nature without the use of synthetics is described separately in Chapter 7.

Procedure for the selection of basic element times

This is as follows—

(*a*) Carry out initial time study programme until there is sufficient data to commence classification of elements into constant and variable categories, and their entry on *summary sheets* on this basis.

(*b*) Allocate set-up numbers for the *summary, graph, data* and other *sheets* used for analysis and compilation of data.

(*c*) Segregate elements into constant and variable categories.

(*d*) Select basic elements of constant length.

(*e*) Select basic elements of variable length.

(*f*) Extend time study programme where necessary.

(*g*) File time studies for reference.

These steps are now explained in more detail.

(*a*) *Carry out initial time study programme*

Since the work being studied may initially be unfamiliar to the observer it is advisable to take a selection of short exploratory studies over a range of work before any attempt is made to summarize the result. Throughout the whole programme it should be ensured that different observers end all similar elements at the same break point to allow equable comparison.

After the very first studies, succeeding ones should increase in length, including at least one long study over several operating cycles, and covering a fair spread of different products if possible. All relevant elements of work should be recorded, including those which occur during work preparation and clearing up operations.

For this reason a few studies should be taken at the extremes of each working period, i.e. when work commences in the morning, after the meal break, just before the midday meal, and at the end of the day or shift.

Care must be taken not to allow too many studies to be taken unnecessarily, and each should be compared with its predecessors to ensure that excessive study work is not being concentrated on one type or part of the operation. The intention at this stage should be to record as much information of a varied character as possible, including all the occasional elements and sufficient repetitive elements necessary to allow segregation of constant from variable elements to commence.

(*b*) *Allocate set-up numbers for the summary and other sheets used*
The set of sheets which are used for analysis and collation of time study data is known as a *set-up*. Before any data is transferred from *work study top sheets* the set-up should be registered on a record of a type shown in Fig 15, the allocated reference number being used to head all the subsequent forms.

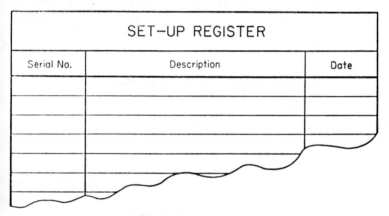

Fig 15 Set-up register

(*c*) *Segregate elements into constant and variable categories*
This must be carried out extremely carefully before any entries are made to the summary sheets. The following method is recommended.

Arrange the time studies in such a way that all the top sheets are visible at the same time. Consider each element in turn on the basis of whether it should logically vary or remain constant, and note whether the recorded values substantiate this. Doubtful time variations should be checked for clerical error or earmarked for further investigation, no element being finally classified until there are valid reasons for so doing.

Sometimes, what is felt should be a constant element may in fact vary because of small differences in design of similar type tools or machines. For example, two elements "switch on machine" may vary because a different type of starting lever is fitted to one of the machines, although they are identical in all other respects. A certain amount of cross-checking on the shop floor may be necessary to clear up such queries.

It is emphasized here that no element which is considered un-necessary should at this stage be marked for non-inclusion in any subsequent analysis. If necessary the method of performing the work should be investigated and further studies taken.

Final selection of average basic times for the two types of element involved two quite different procedures. Each of these is now considered separately.

(d) *Select basic elements of constant length*

(i) Transfer the following information from the *work study top sheet* (Fig 16), to the *study summary—constant elements sheets* (Fig 17) having first headed these with the set-up number and sequentially numbering them—

Study number
Date
Taken by (initials of time study observer)
Operator (operator's name)
NET (effective time from study)
No of cycles (where applicable)
Average rating
Output (quantities and types studied)
Brief description of all elements
Basic element times in centiminutes to the nearest first decimal point
Frequency of occurrence of element times.
The frequency by which elements occurred in the study from which they are transferred is indicated by an index number against the basic time. For example, if element time is 26·2 basic centiminutes and this occurred 18 times during the study, it would be recorded as $26 \cdot 2^{18}$ on the summary sheet and so on (see Fig 17).

(ii) Examine each recorded element to ascertain whether sufficient numbers of observations and studies have been taken to give a satisfactory basis for final selection. This will depend on four main considerations, the total variance of the times themselves, the importance of the element to the operation, the number of observations so far taken, and the nature of the work. Elements which occur frequently on repetitive work, and where it is known that the work content should not logically vary,

WORK STUDY TOP SHEET

Department __Dial gauges.__   Study No. ___101___

Section __Spray shop.__   Date ___4·8·71___

Product __6" diameter dials__   Taken by ___D.W.___

Operation __Spray dials with__   Operator ___L. Jones.___

__white enamel paint__   From 7.59 a.m. to 9.33 a.m.

Estimated Av. Rating __101__

No. of cycles __—__ Output __410__   Elapsed time __94·0 mins__

Plant details __Water wash spray booth.__   Check time __1·3 mins__

Remarks _____   Ineffective time __5·7 mins__

_____   Net effective time __87·0 mins__

Element reference	Element	Basic time centimins.	Frequency	
A	Switch on compressor and stove	180·0	1	
B	Procure plain chisis from stores	137·0	1	
C	Fill can with paint and mix	60·0	1	
D	Charge gun with paint	35·5	2	
E	Procure empty tray from trolley	17·0	18	
F	Pick up dial from box, wipe with cloth, spray with white paint and place aside to tray	17·2	410	
G	Put aside tray of dials to conveyor	26·2	18	
H	Remove grit from dial and re-spray	40·0	1	
I	Procure paint solvent from store	250·0	1	
K	Wash gun with paint solvent	41·0	1	
L	Wash hands	18·0	1	
M	Put aside tray of stoved dials to trolley	6·4	12	
N	Wipe tray clean with solvent on cloth	14·0	1	
P	Push empty trolley to storage bay	41·0	1	
Q	Position full trolley by paint booth	8·0	1	
R	Push trolley of stoved dials to print room	25·0	1	

Fig 16 Work study top sheet being used to summarize a study
on spraying dials with white paint

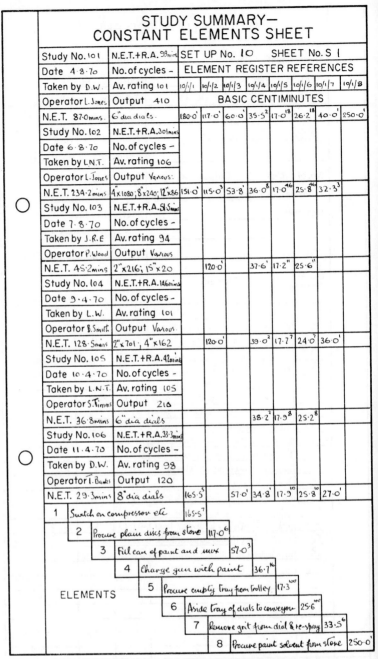

Fig 17 A study summary sheet used to enter elements of constant length from different studies for comparison and selection

should be within $\pm 2\frac{1}{2}$ per cent of each other. On the other hand, a much greater variation can be tolerated on elements which occur infrequently, since their ultimate accuracy has relatively far less influence on the final standard time.

Some frequently occurring elements which cannot be truly classed as variable may still tend to vary widely. Examples of this sometimes happen on operations such as polishing, where unavoidable fluctuations in the surface of the material being polished cause similar fluctuations in basic time. If such conditions are unavoidable, sufficient observations should be taken to allow a reliable average value to be calculated.

(iii) Carefully note those elements for which insufficient data is available. Use these notes to formulate the further time study programme for their specific collection. It may only be necessary to take a small quantity of short studies which concentrate solely on these.

(iv) Obtain average basic times by multiplying selected element times from each study by its frequency, adding these together and dividing by the total of all the frequencies. Values which vary widely in time should be discounted if there is no logical reason for this.

For example, if elemental times under consideration are as follows—

$$16 \cdot 5^{20}; \ 18 \cdot 3^{17}; \ 16 \cdot 2^{31}; \ 16 \cdot 3^{40}; \ 17 \cdot 0^{50};$$

The second of these ($18 \cdot 3^{17}$), is seen to be out of line with the others. Referring back to the original study revealed that notes had been made on it to the effect that it was a "poor study because of a non-cooperative operator". The value was therefore discounted and the selected basic time obtained as follows—

$$\frac{(16 \cdot 5 \times 20) + (16 \cdot 2 \times 31) + (16 \cdot 3 \times 40) + (17 \cdot 0 \times 50)}{20 + 31 + 40 + 50}$$

$$= 16 \cdot 6 \text{ centiminutes}$$

This is a true weighted average of all the selected observations.

(e) *Select basic elements of variable length*

(i) Using a *study summary and register of variable elements*

53

3

sheet for each type of variable element, enter the following information on the head of each—

Department
Section
Product (general description, part number, etc.)
Operation or element description
Set-up number
Sheet number
Plant details

The frequency basis is not entered at this stage.

(ii) Transfer the following details of variable elements from the *work study top sheet* (Fig 16), to the *study summary and register of variable elements sheet* (Fig 18)—

Study number
Date (the date on which the study was taken)
Detail (particulars of size or classification of product etc.). In the *example* Fig 18 the size of the dials was entered under this column
Basic element times in centiminutes to the nearest first decimal point
Frequency of occurrence of element times (expressed as index figures as before).

(iii) Carefully examine each set of variable elements transferred from the top sheets and attempt to discover the reason for their variation.
Satisfactory basis for variation can often be found using a purely logical approach. Consider for example, variable elements recorded during the study of painting operations. Given the same grade of paint and similar surface conditions, basic times would tend to increase with the area painted, and plotting area against time on the graph often produces a straight line—or something very near to it. On the other hand, painting the same area will usually increase the time as the absorption power of the surface increases; in other words, the more porous the surface the longer it will take to paint it. Similarly, different paints may alter the times taken to paint the same surface all other conditions being equal.
Another factor which may influence painting time is complexity of shape. This not only increases the area to be painted, and therefore the time, but it also tends to

STUDY SUMMARY AND REGISTER OF VARIABLE ELEMENTS SHEET

Department __Dial Production__   Set up No. __10__

Section __Spray Shop.__   Sheet No. V __1__

Product __Clock dials__   Frequency Basis __Basic times__

Operation or Element __Pick up dial__   __Vary as the square of the__

__from box, wipe with cloth, spray with__   __diameter of the dial__

__white paint and place aside to tray__   Plant details __Water wash__

Remarks _____   __Spray booth and drying__

_____   __conveyor.__

Ref. No.	Study No.	Date	Details	Basic Centiminutes		% R.A.	% Cent.	Standard minutes per 100 occs.
				Actual	Selected			
1	101	4.8.70	6" dia.	17.2^{410}	17.0	15	3	20.0
2	102	6.8.70	4" dia.	14.1^{1080}	14.4	15	3	17.0
3	102	6.8.70	12" dia.	32.2^{36}	31.5	15	3	37.2
4	102	6.8.70	8" dia.	20.9^{240}	20.6	15	3	24.3
5	103	7.8.70	2" dia.	12.8^{216}	12.9	15	3	15.2
6	103	7.8.70	15" dia.	41.5^{20}	42.1	15	3	49.6
7	104	9.8.70	2" dia.	12.5^{16}	12.9	Ref.	No	5
8	105	10.8.70	6" dia.	17.1^{216}	17.0	Ref.	No	1
9	106	11.8.70	8" dia.	21.4^{144}	20.6	Ref.	No	4
10								
11								
12								
13								
14								
15								
16								
17								
18								
19								
20								
21								
22								
23								
24								
25								

Fig 18 Variable elements of the same character are entered from the study top sheets to this form. After comparison by graphical or other means, selected values are entered alongside and extended to standard times by addition of contingency and relaxation allowances

slow down the operation still further because of the increased difficulty of covering the inaccessible areas. Also, if shapes vary, surface area may vary according to both weight and maximum length, since for objects of the same weight surface area increases in proportion to the increase of the greatest length.

In the *example* of spraying dials, the time for applying the paint was reckoned to become longer as the area of the dial increased. This theory was now tested graphically.

(iv) Graph or list variable elements against variant bases to obtain a definite relationship one with the other. When carrying out this procedure, it is always more convenient to reduce relationships to a straight line form if possible, i.e.

$$y = mx + c$$

where x and y are the variants and m and c are constants. In the *example*, considering the variable element— "Pick up dial from box, wipe with cloth, spray with white enamel and place aside to tray."

This was judged to vary according to the area sprayed, and since—

	Area	$\propto$	time
or	πd^2	$\propto$	time
therefore	d^2	$\propto$	time
where	d	=	diameter of dial

In other words, the square of the diameter was judged to vary directly with the basic element time and accordingly, diameters were plotted on a square base against these times using a *graph sheet* for the purpose. The result was a straight line (see Fig 19).

(v) Determine basic element times. Basic variable element times which had been plotted graphically should be selected directly from the graph and entered to the *study summary and register of variable elements sheets* under the column head "selected basic centiminutes."

Average basic element times classified in groups should be obtained by the selected arithmetic mean method as described under "Select basic elements of constant length (d)" above.

In the *example*, variable element times selected directly from the *graph sheet* (Fig 19) were entered to the study

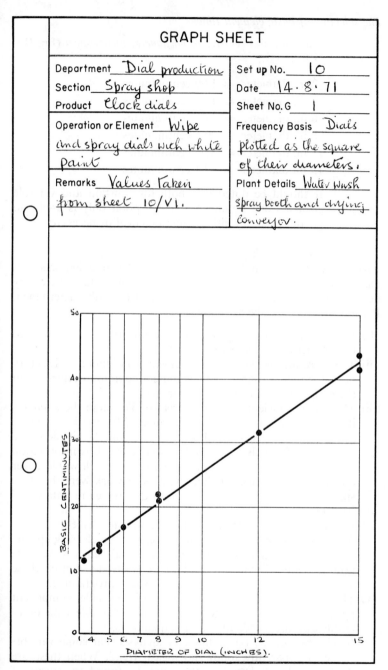

GRAPH SHEET

Department Dial production   Set up No. 10
Section Spray shop   Date 14·8·71
Product Clock dials   Sheet No. G 1
Operation or Element Wipe   Frequency Basis Dials
and spray dials with white   plotted as the square
paint   of their diameters.
Remarks Values taken   Plant Details Water wash
from sheet 10/V1.   spray booth and drying
                               conveyor.

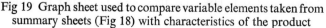

Fig 19 Graph sheet used to compare variable elements taken from
summary sheets (Fig 18) with characteristics of the product

summary and *register of variable elements sheet* (Fig 18), as follows—

  2 in. diameter $=$ 12·9 centiminutes
  4 in. diameter $=$ 14·4 centiminutes
  6 in. diameter $=$ 17·0 centiminutes
  8 in. diameter $=$ 20·6 centiminutes
12 in. diameter $=$ 31·5 centiminutes
15 in. diameter $=$ 42·1 centiminutes

It should be noted that both for constant and variable elements one significant decimal point only is used—no greater accuracy either being practical or necessary.

(*f*) *Extend time study programme where necessary*

After entry of the preliminary study data to the summary sheets, it may be discovered that extra and more specific data is necessary before all the element times can be satisfactorily determined. Reasons for this may be as follows—

(i) Insufficient recordings of certain elements are available to obtain satisfactory selective values.

(ii) There are insufficient variable elements to obtain a satisfactory basis for their variance. This is quite a common occurrence in the early stages of a study programme.

(iii) There is insufficient range of variance. To quote the *example*, it was known that dials of a full range of from 2 in. to 20 in. were required to be sprayed, but data was only available in the early stages for dials of 4 in. to 8 in. diameter. More information was therefore necessary to produce a graph which would cover the whole range. Special short batches of work were therefore specifically arranged to take place so that the extra data could be collected.

(iv) Insufficient or incomplete information on elements of infrequent occurrence such as work preparation, clearing up, etc. If there are found to be insufficient occasional elements available arrangements should be made specifically to collect these elements alone as an extension to the study programme.

(v) The need to break down some elements into smaller component parts in order to separate constant from variable data or to obtain two independently varying quantities.

It may be noticed on preliminary analysis that some variable elements apparently follow no logical trend, When these are further analysed, however, they may be found to consist of either a constant or a variable element or two independent variables, which if segregated can be satisfactorily selected according to recognizable characteristics of the work or product.

(g) *File time studies for reference*

Time studies are expensive documents and should be treated as such. As each element is entered from the *top sheet* to the *summary sheet* it should be carefully ticked off on the *top sheet*, and when all elements have been so entered, the whole of the *top sheet* struck through with a pencil or pen and immediately filed for reference.

5

Relaxation allowances

Fatigue is caused by an accumulation of waste matter in the bodily system due to energy consumption. The cure for this is rest. The amount of energy expended due to the performance of work is roughly proportionate to the exertion of physical and mental effort, although it can also be influenced by the nature of the surroundings. Rest requirements due to fatigue, therefore, tend to increase with both the severity of the environment as well as the physical and mental strains. But operators also need time for personal hygiene. Thus, what is known as relaxation allowance (*RA*) includes both a *fatigue allowance* and a *personal needs allowance*. It can be described as the allowance necessary for an operator to cease work in order to attend to his personal needs and take compensatory rest.

The computation of relaxation allowances

Relaxation allowances are expressed in the form of a percentage which is added to the basic times to form work values. No truly scientific method exists for their computation. They can only be assessed by careful consideration of the nature of the working conditions with a sympathetic understanding of the physiological and psychological limitations of the human body.

All work needs rest, however light it may be. The lowest value is reasonably assessed at 6 per cent. A fair allowance for personal needs is 4 per cent for men and 6½ per cent for women. These give minimum relaxation allowances of 10 per cent and 12½ per cent respectively—equivalent to approximate totals of 43 and 53 minutes in an 8-hour day.

Values in excess of these can best be assessed by detailed study of the working conditions. If the results of several of these studies are analysed, values can be allocated to various different conditions of strain and working environment in such a way that is then possible to compute allowances without the need for further research. The following data is the result of such analyses.

Arrangement of data

The data has been arranged in the following groups, each group being divided into sub-groups for convenience of computation. Final assessment of relaxation allowances is carried out by considering each sub-group separately, allocating values where appropriate, and adding the results together.

1 PHYSICAL STRAINS
   (a) The extent of the average force exerted
   (b) The bodily posture adopted
   (c) The restriction of bodily movement imposed by external working conditions
   (d) The extent of vibration to which the body is subjected
   (e) The shortness of the operation cycle

2 MENTAL STRAINS
   (a) The monotonous nature of the operation
   (b) The degree of concentration necessary
   (c) Eye strain

3 WORKING ENVIRONMENT
   (a) External noise
   (b) Air pollution
   (c) Dirt
   (d) Wet
   (e) Temperature and humidity

4 MINIMUM ALLOWANCES
   (a) Minimum fatigue allowance
   (b) Allowance for personal needs

These are now examined in more detail.

Physical strains

(a) Average force exerted

Consider the average force exerted over the whole element cycle. For example, a force of 20 lb exerted for one-quarter of the period would be taken as—

$$\frac{20}{4} = 5 \text{ lb}$$

for the whole of the element.

61

(i) Negligible = 0–5 lb average force.
This is the physical effort which occurs on such operations as light assembly or clerical work.　　　　　　　　0%–1% *RA*

(ii) Slight = 6–10 lb average force.
This is the slight physical effort which occurs on medium assembly work.　　　　　　　　2%–3% *RA*

(iii) Small = 11–20 lb average force.
This occurs on light hammering, filing, sawing or lifting operations.　　　　　　　　4%–5% *RA*

(iv) Medium = 21–40 lb average force.
This occurs on heavy sawing or filing, medium hammering or lifting operations.　　　　　　　　6%–10% *RA*

(v) Medium heavy = 41–60 lb average force.
This applies to heavy hammering or medium/heavy lifting operations.　　　　　　　　11%–18% *RA*

(vi) Heavy = 61–80 lb average force.
This applies to severe strains caused by lifting heavy loads.
19%–30% *RA*

(vii) Very heavy = 81–100 lb average force.
This applies to very severe strains caused by lifting very heavy loads.　　　　　　　　31%–45% *RA*

(viii) Exceptional = 101–150 lb average force.
This applies to exceptionally heavy or awkward lifting operations.　　　　　　　　46%–80% *RA*

(*b*) *Bodily posture*

Consider the bodily posture which must be adopted to carry out the work.

(i) Sitting.
Work carried out in a sitting posture such as light assembly or clerical work.　　　　　　　　0% *RA*

(ii) Standing.
Light work which must be performed standing or walking freely on level ground.　　　　　　　　1%–2% *RA*

(iii) Confined.
Work which must be performed in a lying position, using a foot pedal while standing, or walking up gradients, stairs, etc.
3%–4% *RA*

(iv) Cramped.
Crouched or stooping position, climbing, etc.
5%–10% *RA*

(*c*) *Restriction of bodily movement*

Consider the amount of restriction placed on the movement of the body when working in confined spaces or wearing protective clothing.

(i) Low.
Low restriction of bodily movement which occurs on operations performed sitting or moving freely. $\qquad$ 0% *RA*

(ii) Medium.
Medium restriction of bodily movement such as occurs when wearing heavy industrial gloves, goggles, etc., carrying slightly unbalanced loads, or using one hand only under load,
$\qquad$ 1%–4% *RA*

(iii) High.
High restriction of bodily movement due to the wearing of heavy protective clothing, e.g. a combination of goggles, rubber gloves, aprons, respirators and thigh boots, etc.; working with hands above head or carrying heavy unbalanced loads.
$\qquad$ 5%–8% *RA*

(iv) Excessive.
Excessive restriction of bodily movement such as might occur when working in confined spaces, e.g. cleaning out ventilator shafts, boiler interiors, etc. $\qquad$ 9%–15% *RA*

(d) Vibration

Consider the amount of vibration to which the body is subjected when performing the operation.

(i) Light.
Vibrational forces which occur on such work as light automatic riveting or chiselling operations. $\qquad$ 0%–2% *RA*

(ii) Heavy.
Heavy vibrational forces which occur on such operations as pneumatic drilling or hammering. $\qquad$ 3%–7% *RA*

(e) Short cycle operation

Consider the amount of shock load to which the body is subjected due to the shortness of the operation cycle.

(i) Light.
Operations of 5–10 seconds' duration (8–16 centiminutes).
$\qquad$ 0%–2% *RA*

(ii) Heavy.
Operations of under 5 seconds (8 centiminutes) duration.
$\qquad$ 3%–7% *RA*

Mental strains

(a) Monotony

Consider the monotonous nature of the operation caused by continuous repetition over a short cycle, difficulty of maintaining interest, or lack of contact with other personnel.

63

(i) Low severity.
Highly repetitive operations where contact is maintained with other workers, or low repetitive operations where contact with others is non-existent.   *0%–1% RA*

(ii) High severity.
Highly repetitive operations where there is little or no contact with others   *2% RA*

(b) *Degree of concentration*

Consider the seriousness of the consequences if the concentration relaxed, and the amount of energy lost through anxiety.

(i) Low.
Where a certain amount of attention to detail is necessary such as occurs on simple inspection operations, simple clerical work, etc.   *0%–2% RA*

(ii) Medium.
Where a fairly high degree of attention to detail is necessary such as might occur on complex inspection operations, fine weaving, spinning, coil winding, precision machining operations, or clerical work involving difficult arithmetic, etc.
3%–4% RA

(iii) High.
Operations involving a high degree of concentration such as complex figure work, or very high precision or dangerous operations where relaxation of concentration may have serious consequences.   *5%–8% RA*

(c) *Eye strain*

Consider the amount of eye strain induced by performing the operation, bearing in mind that this is liable to be greater in poor rather than adequate lighting, or when looking at glaring colours or lights.

(i) Low.
This occurs on operations where attention is occasionally needed in reading numbers on such as adding machines, documents, gauges, meters or clocks in good lighting conditions.   *0%–1% RA*

(ii) Medium.
This occurs on operations where more continuous attention is needed as in precision machining or simple visual inspection.
2%–3% RA

(iii) High.

Occurs where close attention is needed on fine visual inspection or where glaring colours or lights are involved in simple operations. It also occurs where the occasional use of a magnifying glass is necessary.   *4%–5% RA*

(iv) Exceptional.

Occurs where detailed attention is necessary as on inspecting very small and complex detail needing the constant use of a magnifying glass or microscope.   *6%–8% RA*

Assessment in each of the above groups is adequate if the low side of the range is used where lighting and colour conditions are not aggravating, and the higher part of the range is used where the lighting is poor, or glaring colours are present.

Working environment

(*a*) *Noise*

The presence of noise absorbs energy. No allowance should be given if there are normal background noises where speech can be heard without great difficulty. Sudden noises should be given greater allowances than continuous noise of the same volume.

(i) Low severity.

Background noises experienced for example in automatic turret lathe shops or weaving sheds where communication by normal speech is impossible.   *0%–1% RA*

(ii) High severity.

Excessive or sudden noises such as occur on pneumatic drilling or heavy and continuous hammering in boiler making shops, etc.
                    2% RA

(*b*) *Air pollution*

Consider the extent to which the air is polluted and can have an effect on the respiratory organs.

(i) Low.

Occurs where the atmosphere is stuffy or slightly fume-ridden, dusty, etc., as in polishing shops or workshops where chemical fumes are present.   *0%–2% RA*

(ii) High.

Occurs where excessive toxic fumes or dust are present as when cleaning out dust extractors, etc.   *3%–7% RA*

(*c*) *Dirt*

Consider the amount of dirt or other obnoxious material in which it may be necessary to work.

Relaxation allowances

(i) Low.
  Occurs where dust, oil or grease is present which soils the hands or limbs. $0\%-1\% RA$

(ii) High.
  Occurs where excessive dirt such as graphite grease, soot, sewage, etc. are present, causing excessive soiling. $2\%-3\% RA$

(d) Wet

Consider the amount of water present on the hands or surrounding the operation.

(i) Low.
  Working intermittently with hands in wet or on wet floors, such as developing photographic prints or negatives, plating or washing processes, etc. $0\%-1\% RA$

(ii) High.
  Working outdoors in rain or in continuous wet conditions of running water on floor and hands. $2\% RA$

(e) Temperature and humidity

Consider the combined effect on the body of reduced or increased temperature combined with humidity.

(i) Below freezing (below 0°C).
  Working outdoors in conditions below freezing or in a refrigeration plant $10\%-15\% RA$

(ii) Below freezing (below 0°C) as condition (i) but in potentially more humid atmosphere $10\%-15\% RA$

(iii) Cold and dry conditions (over 0°C to 13°C).
  Working outdoors in cold conditions or indoors in non-heated workshops in cold weather but in dry atmosphere.
  $0\%-7\% RA$

(iv) Cold and humid conditions (over 0°C–13°C).
  Working in cold damp conditions. $0\%-10\% RA$

(v) Normal temperature and dry conditions (over 13°C–21°C).
  Working in reasonable temperature under dry conditions.
  $0\%-1\% RA$

(vi) Normal temperature with humid conditions (over 13°C–21°C).
  Working in reasonable temperatures but with a humid atmosphere. $1\%-2\% RA$

(vii) Medium temperature with dry conditions (over 21°C–32°C).
  Working in fairly hot conditions near furnaces or in casting shops but with little or no humidity. $2\%-20\% RA$

(viii) Medium temperature in humid conditions (over 21°C–32°C).
  Working in hot humid conditions. $20\%-40\% RA$

66

(ix) High temperature, dry conditions (over 32°C).
Working in excessive heat continuously such as tending a hot furnace, casting hot iron or steel, etc., but without humidity.
40%–100% RA

(x) High temperature and humid conditions (over 32°C).
Working in excessively hot and humid conditions such as in a laundry drying room or with steam boilers, etc. 100%–150% RA

Minimum relaxation allowances

As previously explained the minimum relaxation allowance is 10 per cent for men, being made up of a personal needs allowance of 4 per cent and a fatigue allowance of 6 per cent. Extra allowances as here described may be added according to the conditions.

Extra allowances for females

An extra personal needs allowance of $2\frac{1}{2}$ per cent should be added for female workers. Statutory regulations exist which restrict women from performing heavy manual work, and an extra $2\frac{1}{2}$ per cent should be added for average force exertion up to 10 lb, and an extra 5 per cent for exertion over this amount.

Relaxation allowance computation sheet

A computation sheet which summarizes the above specifications is shown in Fig 20.

In order to compute an allowance, each category is separately considered and the appropriate percentages entered in the end column on the form and added together, rounding off to the nearest $2\frac{1}{2}$ per cent. It is quite unnecessary to work to any greater degree of accuracy, as relaxation allowance can only be assessed approximately and not determined accurately.

For convenience, it is recommended that a sheet of this nature is produced by a wax stencil or spirit duplicator to allow rapid compilation of allowances.

It is not necessary to compute values for every work element individually. The vast majority of allowances will be found to lie between 10 per cent and 25 per cent. These can be specified in general terms on *data sheets*, reference being made to the relevant relaxation allowance computation sheet. Special computations will then only be necessary for work performed under exceptional circumstances.

Typical examples of RA taken from practice

On page 69 are values of relaxation allowance taken from practice. They apply to male workers.

RELAXATION ALLOWANCE COMPUTATION SHEET

ELEMENT REFERENCES *Examples on Sheets 10V1 and 10F14*			REF No. RA 1	

DESCRIPTION OF ELEMENT *Work elements which are directly exposed to sprayed paint or fumes from paint solvent.*

1(a) *Average force exerted* (i) Light assembly (ii) Med. assembly (iii) Light lifting or hammering (iv) Sawing, filing or med. weight lifting (v) Heavy hammering or med./heavy lifts (vi) Heavy weight lifting (vii) Very heavy weight lifting (viii) Exceptionally heavy lifting	(i) Negligible 0–5 lb	0–1%	(ii) Slight 6–10 lb	2–3%	
	(iii) Small 11–20 lb	4–7%	(iv) Medium 21–40 lb	6–10%	
	(v) Med.-heavy 41–60 lb	11–18%	(vi) Heavy 61–80 lb	19–30%	
	(vii) Very heavy 83–100 lb	31–45%	(viii) Exceptional 101–150 lb	46–80%	
1(b) *Bodily posture*	(i) Sitting	0	(ii) Standing	1–2%	1
	(iii) Confined	3–4%	(iv) Cramped	5–10%	
1(c) *Restriction of bodily movement* (confined spaces or protective clothing)	(i) Low	0	(ii) Medium	1–4%	
	(iii) High	5–8%	(iv) Excessive	9–15%	
1(d) *Vibration*	(i) Light	0–2%	(ii) Heavy	3–7%	
1(e) *Short cycle operation*	(i) 5–10 secs	0–1%	(ii) Under 5 secs	2–3%	
2(a) *Monotony*	(i) Low	0–1%	(ii) High	2%	
2(b) *Degree of concentration*	(i) Low	0–2%	(ii) Medium	3–4%	1
	(iii) High	5–8%			
2(c) *Eye strain* (due to close attention)—Low, good light— High, poor light	(i) Low	0–1%	(ii) Medium	2–3%	
	(iii) High	4–5%	(iv) Excessive	6–8%	
3(a) *Noise*	(i) Low	0–1%	(ii) High	2%	
3(b) *Air pollution*	(i) Low	0–2%	(ii) High	3–7%	1
3(c) *Dirt*	(i) Low	0–1%	(ii) High	2–3%	1
3(d) *Wet*	(i) Low	0–1%	(ii) High	2%	1
3(e) *Temperature* Below freezing	(i) Dry	15–10%	(ii) Humid	15–10%	
Cold 0°C–13°C	(iii) Dry	7–0%	(iv) Humid	10–0%	
Normal 13°C–21°C	(v) Dry	0–1%	(vi) Humid	1–2%	
Medium 21°C–32°C	(vii) Dry	2–20%	(viii) Humid	20–40%	
High 32°C upwards	(ix) Dry	40–100%	(x) Humid	100–150%	
4 *Minimum relaxation allowance*					10
(Round off to nearest 2½%)	**Total relaxation allowance (men)**				15%
Add 2½% for lifts up to 10 lb Add 5% for lifts over 10 lb	**Total relaxation allowance (women)**				17½%

Fig 20 Relaxation allowance computation sheet (all allowances include "tea breaks")

15% Applies to operations performed standing demanding some physical effort or which are monotonous. It also applies to other operations which need concentration or which are to some degree performed in a slightly disturbing environment such as noise, dirt, air pollution, etc.

25% Applies to medium/heavy work or work performed in somewhat trying conditions such as excessive fumes, dirt or under slightly higher temperature conditions than normal.

40%–60% Applies to operations which are encountered in iron and steel works where excessive amounts of physical energy are needed in extreme conditions of heat and sometimes humidity.

Outdoor work

In those cases where work has to be performed out-of-doors in adverse weather conditions, extra allowances may have to be given to operators who are on bonus incentive. The data will provide a guidance for so doing. If, however, such weather conditions lead to a total stoppage of work, this should be classed as lost time (see Chapter 14).

Relaxation allowances used on example

Allowances used for the example of "spraying dials with white enamel" are given below. These were first computed using the standard sheet, and then entered to a *data sheet,* for eventual use against elements on the *register of constant elements sheet* and *study summary and register of variable elements sheet.*

The operations were being performed by male operators—

$12\frac{1}{2}$% Light operations performed in the spray shop which do not involve excessive dirt or direct exposure to the sprayed paint.

15% Operations performed in the spray shop which involve slight physical effort but which do not involve excessive dirt or direct exposure to the sprayed paint. Light operations performed in the spray shop where the hands are likely to become soiled, and operations which are exposed to sprayed paint or fumes from paint solvent.

The *relaxation allowance computation sheet* (Fig 20) is shown completed for the 15 per cent allowance above.

Tea breaks

Observations made in the past on workers over prolonged intervals, indicated that relaxation is more beneficial when the greater part of

Relaxation allowances

it is taken as a definite break in the work of some 10 minutes or so, rather than in a series of shorter pauses. This led to the introduction of "compulsory rest periods" in certain establishments on an experimental basis. The idea caught on—particularly during the last war—when tea and other refreshments began to be provided. Thus the tea break came into being. It is part of the relaxation allowance, and values computed by the data given in this chapter are inclusive of this allowance.

6

The compilation of basic synthetic data

Synthetic data usually consists of elemental operations and frequencies which can be combined in various ways to compute standard times for complete operations.

The following procedure can be used to develop such data whatever the nature of the work being measured.

Procedure

(*a*) Transfer selected basic times for constant elements from summary sheets to *register of constant elements sheets*.

(*b*) Add relaxation allowances against elements on the element registers.

(*c*) Classify elements according to the basis of their origin.

(*d*) Determine the contingency allowance.

(*e*) Enter the contingency allowance against all appropriate elements on the registers.

(*f*) Calculate elemental standard times.

(*g*) Collect together all elements to common bases of origin.

(*h*) Further collect together groups of elements to form the minimum number of elemental operations.

(*j*) Determine the frequencies of variation of elemental operations to the main basis of the final standard times.

This is now explained in more detail.

The compilation of basic synthetic data

(*a*) *Transfer selected basic times to the register of constant elements sheets*

This should be carried out as follows—

(i) Critically examine each element entered to the *study summary— constant elements sheets* to determine whether it really contributes to the final operation in its present form, i.e. whether it can be reduced in size or eliminated altogether by rearrangement of the method of working. If during this examination, it is felt that there is scope for improvement in this way, practical experiments should be carried out and studied on the shop floor and revised elemental times obtained under the new conditions before proceeding further.

(ii) Transfer all selected basic elemental times considered to be valid from the *study summary—constant elements sheets* to the *register of constant elements sheets*, together with the element descriptions. Each element so transferred should be cross-referenced back by entering the originating *summary sheet* and element number in the column headed "Summary reference". For example, Element No 6 from Sheet No S1 on set-up No 10 is referred to as Summary reference 10/S1/6 and so on (see Figs 17 and 21).

Transferred elements should also be cross-referenced forwards by similarly entering the element register references at the tops of the columns so headed, Element No 8 on *register of constant elements sheet* No 1, set-up No 10, being referred to as 10/1/8 and so on (see Fig 17).

(iii) Head the *register of constant elements sheets* (Fig 21) with the appropriate details of the department, section, product, operation, date, plant details, etc., the "type of set-up" will always be "synthetic" unless a standard time for operations of a unique character is being computed (see Chapter 7).

Although the procedure at this stage may seem tedious, it will be found to be invaluable when elements need to be traced to their originating studies during extension of the scope of synthetic data, or where arithmetical errors have been made.

(*b*) *Add relaxation allowances to the element registers*

This consists of using relaxation allowances which have been computed and entered to *data sheets* as explained in Chapter 5 and adding these against the appropriate elements on the *register of constant elements sheets* and the *study summary and register of variable elements sheets* (see Figs 18 and 21).

REGISTER OF CONSTANT ELEMENTS SHEET

Department _Dial production_
Section _Spray shop._
Product _Clock dials_
Operation _Spray dials with white enamel paint and stove_
Remarks _____

Set up No. _10_
Date _12·8·71_
Sheet No. _1_ of _1_
Type of set-up _Synthetic_
Plant details _Water wash spray booth and drying conveyor._

Ref. No.	Summary ref.	Element	Selected basic mins. per 100	% R.A.	% Cent	Standard minutes per 100 occs	Freq. basis	Std. mins. per 100
1	10 S1 1	Switch on compressor etc	165·5	12½	–	186·0	Day	
2	10 S1 2	Procure plain discs from store	117·0	12½	–	131·5	Batch	
3	10 S1 3	Fill can with paint and mix	57·0	12½	2½	65·6	Charge	
4	10 S1 4	Charge gun with paint	36·7	15	2½	43·1	Charge	
5	10 S1 5	Procure empty tray from trolley	17·3	12½	2½	19·9	Tray	
6	10 S1 6	Aside tray of dials to conveyor	25·6	12½	2½	29·4	Tray	
7	10 S1 7	Remove grit from dial & re-spray	33·5	15	2½	39·4	Dial	
8	10 S1 8	Procure paint solvent from store	250·0	12½	2½	287·0	Charge	
9	10 S2 1	Wash gun with paint solvent	41·2	15	2½	48·5	Charge	
10	10 S2 2	Aside tray of stoved dials to trolley	6·5	12½	2½	7·5	Tray	
11	10 S2 3	Wipe tray clean with solvent on cloth	13·4	12½	2½	15·4	Tray	
12	10 S2 4	Push empty trolley to storage bay	42·0	15	2½	49·5	Trolley	
13	10 S2 5	Position full trolley of trays by booth	8·3	15	12½	9·8	Trolley	
14	10 S2 6	Wash hands.	16·6	12½	–	18·7	Day	
15	10 S2 7	Push trolley of stoved dials to print room	25·0	15	2½	29·4	Trolley	
16	10 S2 8	Procure drum of paint from stores	481·0	17½	2½	577·0	Charge	
17	10 S3 1	Filter paint through muslin	152·0	15	2½	179·0	Charge	
18	10 S3 2	Dispose of packing paper	26·5	12½	–	29·8	Batch	
19	10 S3 3	Book work on time sheet	32·0	12½	–	36·0	Batch	
20	10 S3 4	Switch off compressor etc	145·5	12½	–	163·5	Day	
21								
22								
23								
24								
25								
					TOTALS			

WORK VALUE IN STANDARD MINUTES PER
COMPILED BY

Fig 21 Register of constant elements sheet being used to list selected basic times of constant length which are then extended to standard element times by addition of contingency and relaxation allowances, and examined and classified by frequency basis

(c) Classify elements according to the basis of their origin

All work elements have a basic reason for being performed, and the frequency of their occurrence will depend on this basis. For example, an element "move box of parts from bench to floor" exists because boxes are used to contain parts, the frequency of its performance with respect to the parts depending on the number of them in the box. Similarly, if the boxes are subsequently placed on a trolley, then those work elements which are necessary to move trolleys from one place to another will occur with a frequency according to the number of boxes which can be loaded to it. In other words, if ten boxes are placed on each trolley, then trolleys need to be moved with a frequency of one in ten boxes, etc.

Elements are classified in this way by entering the bases of classification under the heading "Frequency basis" on the *register of constant elements sheet* or the *study summary and register of variable elements sheet* for constant and variable elements respectively.

In the *example*, classification was carried out as follows—

(i) Elements per dial.
  These were as follows, their existence depending on the fact that there were dials which needed to be sprayed—
  10/V1 Procure dial, wipe with rag, spray with white paint and lay aside on tray (see Fig 18).
  10/1/7 Remove grit from dial and respray (this occurs when a piece of grit from the atmosphere occasionally settled on a newly painted dial) (see Fig 21).

(ii) Elements per tray (Fig 21).
  These were all concerned with the handling of trays and were constant in length.
  10/1/5 Procure empty tray from storage trolley.
  10/1/6 Aside full tray of dials to conveyor.
  10/1/10 Aside tray of stoved dials to trolley.
  10/1/11 Occasionally wipe surplus paint from tray with cloth dipped in paint solvent.

(iii) Elements per trolley (Fig 21).
  All these were constant elements and referred to the handling of trolleys.
  10/1/12 Move empty trolley from spray booth to stoving position to prepare for unloading.
  10/1/13 Move trolley of empty trays nearer to spray booth.
  10/1/15 Move full trolley of stoved dials to print room.

(iv) Elements per charge of spray gun (Fig 21).
These depended on the number of times that the spray gun was charged with paint.
10/1/3 Fill small can from paint drum and mix paint.
10/1/4 Charge gun with paint.
10/1/8 Procure paint solvent from stores (this was used to wash the gun occasionally).
10/1/9 Wash gun with paint solvent.

(v) Elements per batch of work performed (Fig 21).
These depended on the number of batches of work done by the operator during the day.
10/1/2 Procure plain dials from stores.
10/1/18 Dispose of packing paper in which batches of plain unpainted dials were wrapped.
10/1/19 Book the quantities produced on a work sheet at the end of each batch.

(vi) Elements per working day (Fig 21).
These were performed with respect to each working day.
10/1/1 Switch on the spray paint compressor and the stove. This was performed once in the morning and again after the midday meal.
10/1/20 Switch off the compressor and stove. This occurred once before the midday meal and again before leaving at the end of the day.

(*d*) *Determine the contingency allowance*

A contingency allowance is a collection of elements which is likely to occur legitimately but sometimes infrequently and which when added together does not usually amount to more than about 5 per cent of the work cycle. The method of obtaining this allowance is—

(i) Examine each element which occurs per working day—or other elements that can be so classified—and obtain the standard time per occurrence for each by adding the relaxation allowance to the selected basic time.
In the *example*, all those elements which had been classified as "per day" and "per batch" fell into this category, since the number of batches of work per day were small and more or less constant, and the total time per batch was of very small duration. If, however, the batches of work would have occurred more often and times were of longer duration, it would have been necessary to have calculated a separate standard time for "preparation time per batch of work". Standard times per

75

element were next obtained by multiplying the basic element times by a factor which included the relaxation allowance as follows—

$$v_1 = \frac{(100 + RA)b}{100}$$

Where v_1 = the standard time per element
RA = the relaxation allowance percentage
b = the basic element time

For example, if the selected basic time was 26·5 centiminutes and the relaxation allowance is $12\frac{1}{2}$ per cent, then the standard time would be—

$$v_1 = \frac{(100 + 12\frac{1}{2})26\cdot5}{100}$$

$$= 29\cdot8 \text{ centiminutes per occurrence.}$$

(ii) Enter all these elements together with their standard times to *collection of elements sheets* (Fig. 22).

(iii) Determine the frequency by which each element occurs per working day.
This is either obtained by reference to the *study summary—constant elements sheets* or from local knowledge. In the *example* for instance, since it was known that the paint spray compressor and stove had to be switched on and off twice a day, the frequencies of these elements were expressed as 2/1. Also, from the *summary sheets* it was noticed that the frequency of the change-over elements per batch occurred approximately 3 times per day, and this checked with the average batch size being fed into the workshops. The frequency of these elements was therefore expressed as 3/1 (see *collection of elements sheet* Fig 22).

(iv) Convert standard times per occurrence to standard times per day or shift by multiplying them by the frequencies per day (or shift) and adding the results together.
In the *example*, the gross standard time per day was evaluated as 1308 centiminutes (see Fig 22).

(v) Determine the contingency allowance percentage to the nearest whole number from—

$$\% \text{ Contingency} = \frac{100d}{D-d}$$

Where d = the total standard time per day or shift
D = the total actual time per day or shift.

COLLECTION OF ELEMENTS SHEET

Department _Dial production_   Set up No.   10

Section _Spray shop._   Date _12 . 8 . 71_

Product details _Spray clock_   Sheet No: C   1

dials with white enamel paint   Frequency Basis

Type of Elements _Constant_   _Various_

Remarks _Determination of contingency_   Graph Sheet references

and collection of constant elements

together in groups.

Element reference	Element	Standard minutes per 100 occs	Freq.	Standard minutes per 100 units	Other variables			Total std. mins. per 100 units
	CONTINGENCY ELEMENTS (PER DAY)							
10/1/1	Switch on compressor etc.	186·0	2/1	372·0				
10/1/2	Procure discs from stores	131·5	3/1	394·5	% Contingency.			
10/1/4	Wash hands.	18·7	9/10	17·1	= $\frac{100 \times 13·08}{480-13·08}$			
10/1/18	Dispose of packing paper	29·8	3/1	89·4				
10/1/19	Book work on time sheet	36·0	3/1	108·0	= 2·8%			
10/1/20	Switch off compressor etc.	163·5	2/1	327·0				
	TOTAL PER DAY			1308·0	SAY 3%			
	ELEMENTS PER TRAY							
10/1/5	Procure empty tray	19·9	1/1	19·9				
10/1/6	Aside tray to conveyor	29·4	1/1	29·4				
10/1/10	Aside tray of disks to trolley	7·5	1/1	7·5				
10/1/11	Wipe tray with cloth	15·4	6/100	0·9				
	TOTAL PER TRAY			57·7				
	CONSTANT ELEMENTS PER DIAL							
10/1/7	Remove grit from dial and re-spray	} 39·4	1/210	0·2				
	TOTAL PER DIAL			0·2				

Fig 22 Collection of elements sheet being used to collect and add together constant elements which have the same frequency basis

In the *example—*

$$d = 1308 \text{ standard centiminutes per day}$$
$$= 13{\cdot}08 \text{ standard minutes per day}$$
$$\text{and } D = 8 \text{ hours per day (a normal working day)}$$
$$= 480 \text{ minutes}$$

In this case therefore,

$$\% \text{ Contingency} = \frac{100 \times 13{\cdot}08}{480 - 13{\cdot}08}$$
$$= 2{\cdot}8\%$$
$$\text{say } 3\%$$

Note: Contingency percentages are usually rounded off upwards, i.e. even 2·1 per cent is rounded up to 3 per cent, etc.

(*e*) *Enter the contingency allowance against all appropriate elements on the registers*
This consists of entering the percentage contingency allowance against all elements (except the contingency elements themselves) on the *register of constant element sheets*, and the *study summary*, and *register of variable element sheets* (see Figs 18 and 21).

(*f*) *Calculate elemental standard times*
Standard times for all the constant and variable elements (except the contingency elements) are now calculated from the following approximate formula and entered to the *register sheets* (Figs 18 and 21)—

$$v_2 = \frac{(100 + RA + \text{Contingency})b}{100}$$

Where v_2 = the standard time per element including contingency allowance.
RA = the relaxation allowance percentage.
Contingency = the contingency allowance percentage.
b = the basic element time.

For example, if the selected basic time is 31·5 centiminutes, the relaxation allowance was 15 per cent, and the contingency allowance 3 per cent, then the standard centiminutes per occurrence for the element would be—

$$v_2 = \frac{(100 + 15 + 3)31 \cdot 5}{100}$$

$$= 37 \cdot 2 \text{ standard centiminutes per occurrence}$$

It will be noticed that the formula used is not strictly accurate, as it should be—

$$v_2 = \frac{(100 + RA)(100 + \text{Contingency})b}{100 \times 100}$$

The difference is so small, however, that the simpler formula is adequate.

(g) *Collect together all elements to common bases of origin*
(i) Elements classified during step (*c*) are first entered in groups on the *collection of elements sheets* (Fig 22).
In the *example*, these were now—

Variable elements per dial.
Constant elements per dial.
Constant elements per tray.
Constant elements per trolley.
Constant elements per charge of paint in spray gun.

All such entries are carefully cross-referenced back to their originating *summary sheets* (see Fig 22).

(ii) Against each element is entered the frequency it occurs to the common basis. Consult summary sheets for guidance.
To understand this, consider an element "wipe tray clean with cloth". This was recorded on the *study summary—constant elements sheets* (similar to Fig 17), as being observed a total of 6 times, whereas elements such as "procure empty tray", "aside full tray", etc., occurred 100 times.
In other words, the element occurred with a frequency of 6 out of 100 trays handled or 6/100.
Most elements can be dealt with in this way, basing them on the frequency by which they actually occurred over the full range of studies taken.

(iii) Standard times for the common bases are then calculated by multiplying elemental standard times by their frequencies and adding the results together.
An example of one of these is shown in Fig 22, for elements per tray—

Element description	Standard centiminutes per occurrence	Frequency per tray	Standard centiminutes per tray
Procure empty tray from trolley	19·9	1/1	19·9
Aside full tray of dials to conveyor	29·4	1/1	29·4
Aside tray of stoved dials to trolley	7·5	1/1	7·5
Wipe tray clean with cloth	15·4	6/100	0·9
Total constant elements per tray (standard centiminutes)			57·7

(h) Further collect together groups of elements

Sometimes, when bases of collection are examined, it will be found possible to reduce their number by combining some of them together. For example, if there are "elements per piece" and "elements per box", and the number of pieces in a box is always the same, then the whole can be expressed "per piece" by relating the frequency of the one set of elements to the other and adding the two together.

In the *example*, the following were treated in this way—

(i) Since 10 trays could always be loaded to the trolley, elements per tray and per trolley were combined as follows—

Element description	Standard centiminutes per occurrence	Frequency per tray	Standard centiminutes per tray
Constant elements per tray	57·7	1/1	57·7
Constant elements per trolley	88·0	1/10	8·8
Total constant elements per tray (standard centiminutes)			66·5

(ii) The constant elements per dial (i.e. "remove grit from dial and re-spray"), having been multiplied by its frequency of 1/210 dials was added to the variable elements per dial to give total standard centiminutes per dial (see also Figs 22 and 23)—

COLLECTION OF ELEMENTS SHEET

Department _Dial Production_	Set up No. _10_
Section _Spray shop_	Date _12 · 8 · 71_
Product details _Spraying dials_	Sheet No. C _3_
with white enamel paint	Frequency Basis _Standard mins_
Type of Elements _Variable_	_vary directly as the square of the_
plus constant (10/c/1) per dial	_diameter of the dial._
Remarks _Main element is "pick up_	Graph Sheet references ____
dial from box, wipe with cloth, spray	_10 G3 (variable elements)_
with white paint and aside to tray "	

Element reference	Element	Standard minutes per 100 occs	Freq.	Standard minutes per 100 units	Other variables % /1			Total std. mins. per 100 units
10/vi/5	2" dial	15.2	1/1	15.2	0.2			15.4
10/vi/2	4" dial	17.0	1/1	17.0	0.2			17.2
10/vi/1	6" dial	20.0	1/1	20.0	0.2			20.2
10/vi/4	8" dial	24.3	1/1	24.3	0.2			24.5
10/vi/7	12" dial	37.2	1/1	37.2	0.2			37.4
10/vi/6	15" dial	49.6	1/1	49.6	0.2			49.8

Fig 23 Collection of elements sheet being used to collect and add together variable and constant elements which have the same frequency basis

The compilation of basic synthetic data

	Standard centiminutes per dial		
Dial size	*variable*	*constant*	*total*
2″	15·2	0·2	15·4
4″	17·0	0·2	17·2
6″	20·0	0·2	20·2
8″	24·3	0·2	24·5
12″	37·2	0·2	37·4
15″	49·6	0 2	49·8

(*j*) *Determine the frequencies of variation of elemental operations*
This is carried out in two stages—

(i) Decide the most convenient output unit in which the final standard times are to be expressed. Examples are—

Standard minutes per piece
Standard minutes per 100 pieces
Standard hours per 1,000 pieces
Standard minutes per unit weight (lb, kg, etc.)
Standard minutes per unit volume (gallon, litre, etc.)
Standard minutes per operation cycle (i.e. load, machine or process cycle, etc.)

If it is decided to express output as machine utilization percentage, net standard times must first be obtained per unit output (see Chapter 10).
In the *example*, it was decided to use standard minutes per 100 pieces (dials).

(ii) Establish the frequency of variation of the elemental operation to the main basis of the standard times (i.e. the frequency per piece, per lb, per kg, per gallon, per litre, etc.)
This relationship is a function of the process and is obtained by study of the operation. In the majority of cases it can simply be determined by examination of the study summary sheets and time studies, but in those cases where complex technical processes are involved (such as die casting, plastic injection moulding, metal plating, etc.) it may be necessary to consult specialized technicians, etc.
In the *example*, since standard times were expressed alternatively as "per dial", "per tray" and "per charge of spray gun" it was necessary to establish the relationship of—
    trays to dials
    charges of spray gun to dials.

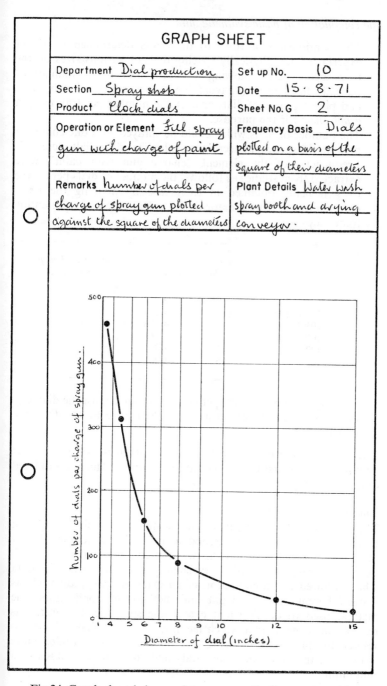

Fig 24 Graph sheet being used to compare the frequencies with
which paint is charged to a spray gun with various diameters of
dials sprayed

The relationship of trays to dials was determined by direct experiment, i.e. physically spacing dials of various sizes on the trays being used, allowing a contingency for a number of partially filled trays (unavoidable at the end of a batch, etc.), and plotting sizes of dials against the number per tray using a *graph sheet* for the purpose.

The number of dials which could be sprayed with one charge of paint in the spray gun decreased as the size of the dials increased. Frequencies of the element "charge gun with paint" were extracted from studies as follows, the number of dials per charge of gun being obtained in each case by dividing the numerator into the denominator of the vulgar fraction representing frequency—

Study numbers	Diameter of dial	Frequency	Number of dials per charge of gun
103 & 104	2″	2/917	459
102 & 104	4″	4/1242	311
101 & 105	6″	4/626	157
102 & 106	8″	4/360	90
102	12″	1/36	36
103	15″	1/20	20

Plotting numbers of dials per charge of gun against the diameter of the dials in inches on a *graph sheet* gave a smooth curve (see Fig 24).

Before any of the data developed at this stage is expressed in its final form, it has first to be checked for accuracy. This is considered in the next chapter.

Note on the example

The example of "spraying dials with white enamel paint" is quoted to aid explanation of the method of compiling synthetics, but it must not be concluded that this was the complete data developed. On the contrary, it formed a very small part of comprehensive data which allowed the computation of an extensive range of products using different kinds of paint and methods of application. The method used to develop and extend the synthetic, however, was exactly the same as that which has now been described, the cross-referencing procedure ensuring that no element was studied more than once, thus reducing time study and subsequent analysis to a minimum.

7

Checking the accuracy of measured work

The computation of complete standard times per operation

As previously explained, standard times for complete operations can be computed using either of two methods—

(a) From synthetic data.
(b) Directly from data collected by time study or other means.

These are now considered in more detail.

(a) The computation of standard times from synthetic data

At this stage, synthetic data will probably consist of some or all of the following—

(i) Elemental operations of constant duration which apply to a range of products.
(ii) Elemental operations of constant duration which apply to a more restricted range of products than (i) above.
(iii) Elemental operations which vary according to some recognizable characteristic of the products or processes.
(iv) Data which gives frequencies of occurrence of some of the constant elemental operations to the main operation, and which apply to a fairly wide range of products.
(v) Data which gives frequencies of occurrence, but which applies to a more restricted range of products than (iv).
(vi) Variable frequency data in graphical form.

Checking the accuracy of measured work

Standard times for complete operations (either which are currently being carried out or have been studied) should be computed from this data so that it can be checked.

In the *example*, the data at this point consisted of—

  (i) A constant length elemental operation "elements per tray" of 66·5 standard centiminutes (Fig 25).

 (ii) A constant length elemental operation "elements per charge of paint in spray gun" of 55·6 standard centiminutes (Fig 25).

(iii) Variable elemental operations "per dial" with respect to the diameter of the dial (Fig 25).

(iv) Variable frequencies "per dial" for the constant elemental operation "elements per tray" (Fig 25).

 (v) Variable frequencies "per dial" for the constant elemental operation "elements per charge of paint in spray gun" (Fig. 25).

Computing standard times from this data for the complete operation of spraying dials with white enamel was carried out as follows. Values have been inserted for a 6 in. diameter dial to aid the explanation—

Description	Value	Frequency	Taken from	Standard centiminutes per dial
Elements per dial	variable	1/1	graph sheet	20·2
Elements per tray	66·5	1/24	Frequency from graph sheet	2·8
Elements per charge of spray gun	55·6	1/156	Frequency from graph sheet	0·4
			Total	23·4

Rounding off to the nearest $\pm 2\frac{1}{2}$ per cent, this was—

    23 standard minutes per 100 dials sprayed.

(b) Standard times obtained directly from time study data

This procedure is only useful where the operations performed are of a unique nature, i.e. where there is no basis for comparison with any other work being performed within the organization. Instances

86

DATA SHEET

Description **Collection of basic** Set-up No. **10**
synthetic data Sheet No. D **2**
Operation **Spray dials with white paint** Date **17.8.71**
Remarks **Spray shop - Dial production department**

CONSTANT VALUE DATA	Reference
CONTINGENCY ALLOWANCE = 3%	10C1
ELEMENTS PER TRAY = 66.5 Std centimins	10C2
ELEMENTS PER CHARGE OF PAINT TO SPRAY GUN = 55.6 Std. centiminutes	10C2

VARIABLE DATA

Dia. of dial	Elements per dial	Number of dials per tray	Number of dials per charge of spray gun.
	Ref. 10G4	Ref. 10G3	Ref. 10G2
2"	15.4 std centiminutes.	216	460
3"	16.1 std centiminutes.	96	333
4"	17.2 std centiminutes.	54	310
5"	18.5 std centiminutes.	35	232
6"	20.2 std centiminutes.	24	156
7"	22.0 std centiminutes.	15	142
8"	24.5 std. centiminutes.	12	122
9"	26.8 std centiminutes	8	71
10"	29.7 std centiminutes.	6	56
12"	37.4 std centiminutes.	6	36
15"	49.8 std centiminutes	2	20

note: interpolate intermediate values

from appropriate graph sheets.

Fig 25 Data sheet being used to collect basic synthetic data together

of this are rare, and would only apply to single operations such as cleaning or setting a piece of equipment where there is only one method of so doing. Use the *register of constant elements sheets* from stage 3 onwards (see Fig 26)—

(i) Time study the operation as described in Chapter 3.

(ii) Enter basic times from these studies to the *study summary— constant elements sheets* and select basic element times as described in Chapter 4 (*d*). These elements will all be of constant length since there is no basis for comparison.

(iii) Transfer the selected basic times to the *register of constant elements sheet*.

(iv) Compute relaxation allowances as described in Chapter 5.

(v) Add relaxation allowances to all elements on the *register of constant elements sheet*.

(vi) Extend all values with relaxation allowances added, entering these under the column "Standard minutes per 100 occs". Use the following formula—

$$v = \frac{(100 + RA)b}{100}$$

where v = the standard time per element
RA = the relaxation allowance percentage
b = the basic element time

(vii) Enter the frequency which each element occurs to the main basis of the standard time (i.e. per operation, per piece, etc.) in the column headed "frequency basis".

(viii) Calculate the standard centiminutes per operation by multiplying frequencies by the standard centiminutes per occurrence, entering these values in the column which is headed "Std. mins. per 100. . . ."

(ix) Obtain the total standard time by addition of the values in the column which is headed "Std. mins. per 100 . . .", and round this off to the nearest $\pm 2\frac{1}{2}$ per cent for publication.

Checking standard times

Standard or allowed times should be carefully checked against a measured time interval before issue, particularly if they have been built up from synthetic data. This can be achieved in three ways—

1 Checking against existing time studies (known as *check studies* in this context).

2 Checking against specially taken time studies of fairly long duration known as *production studies*.

3 Checking against data gathered by rated activity sampling.

REGISTER OF CONSTANT ELEMENTS SHEET

Department _Dial production_   Set up No. _10_

Section _Spray shop._   Date _20·8·71_

Product _Spray booth_   Sheet No. _7_ of _7_

Operation _Clean out_   Type of set-up _Individual_

Spray booth   Plant details _Water-wash_

   Spray booth fitted with

Remarks _____   _coke filters_

Ref. No.	Summary ref.	Element	Selected basic mins. per 100	% R.A.	% Cent	Standard minutes per 100 occs	Freq. basis	Std. mins. per 100
1	10/59/1	Dress in protective clothing	164	15		189	1/1	189
2	10/59/2	Remove filters, clean, fill	7812	25		9760	1/1	9760
3		with fresh coke						
4	10/59/3	Clean waterway plate	1363	20		1638	1/1	1638
5	10/59/4	Open waste water cock	117	15		135	1/1	135
6	10/59/5	Remove & clean fan baffles.	3173	20		3810	1/1	3810
7	10/59/6	Clean waste pipe filter	762	20		915	1/1	915
8	10/59/7	Clean out paint waste	5062	25		6340	1/1	6340
9		from base of booth						
10	10/59/8	Clean rear & front of booth	1173	15		1350	1/1	1350
11	10/59/9	Take waste to rubbish dump.	1062	15		1225	1/1	1225
12	10/59/10	Replace filters, centre	2163	15		2490	1/1	2490
13		partition & fan baffles						
14	10/59/11	Clean area around booth	735	12½		827	1/1	827
15	10/59/12	Wash hands	27	12½		30	1/1	30
16	10/59/13	Doff protective clothing	173	15		199	1/1	199
17								
18								
19								
20								
21								
22								
23								
24								
25								
					TOTALS			28908

WORK VALUE IN STANDARD MINUTES PER   OCC   **290**

COMPILED BY   _J. Smith._

Fig 26 Register of constant elements sheet being used to compute standard times for work of a unique character, i.e. where there is no basis for comparison with any other work being carried out in the organization

Checking the accuracy of measured work

By far the most reliable method is the use of *production studies* (or a sufficient number of existing studies which are of equal value). Checking by rated activity sampling is more suitable when dealing with the less accurate standards obtained by broad work measurement as described in Chapter 8.

The procedure for checking from time studies (i.e. either check studies or production studies), can be set out as follows—

(*a*) Select time studies already in existence or take special studies for checking purposes.

(*b*) Partially summarize the studies.

(*c*) Check standard times.

(*d*) Further summarize the studies and adjust standard times where necessary.

(*e*) In special cases, check relaxation allowances.

To aid explanation of this procedure, the following example of a production study is incorporated in the text. Figures added from time to time refer to values taken from it—

The study commenced at 8.01 am and finished at 1.00 pm.

The *elapsed time* of the study was therefore 4 hours 59 minutes or 299 minutes.

The total *check times* were 129 centiminutes or 1·29 minutes.

The *study time*, which is the *elapsed time* less the *check time* was therefore 297·71 minutes.

The total *waiting time* was 4·4 minutes.

The total *idle time* was 24·0 minutes.

The *effective time*, which is the *study time* minus the sum of the *waiting time* and the *idle time* was 269·21 minutes.

The *average rating* was 106.

The *output* during the study was—

1080 dials sprayed with paint 4 in. dia.

240 dials sprayed with paint 8 in. dia.

86 dials sprayed with paint 12 in. dia.

The computed standard times for the above dials was—

4 in. dia. dials sprayed with paint = 19 SMs per 100

8 in. dia. dials sprayed with paint = 31 SMs per 100

12 in. dia. dials sprayed with paint = 50 SMs per 100.

The procedure is now considered in more detail.

(*a*) *Select time studies already in existence or take special studies for checking purposes*

Time studies already in existence can be used for checking purposes if—

(i) They are long enough to cover a fair spread and quantity of work.

(ii) There have been no changes in the method of working since they were taken.

(iii) It is not felt necessary to check work not previously studied, but for which standard times can be computed from synthetic data.

(iv) They are adequate to prove the accuracy of standard times which are in dispute.

Since it is important that such studies contain sufficient occasional elements and contingencies, it is sometimes necessary to consider using more than one of them added together if the necessity of taking special studies is to be avoided. This is why it is economical to include studies of this nature in the initial time study programme.

The procedure for taking production studies is exactly the same as that used for any other studies (i.e. as given in Chapter 3) each element being timed and rated at each delay interval accounted for. The practice of rating every half minute or so without recording any other detail is not to be recommended, since if there are discrepancies in the standard times, there is no way of analysing the data to discover the reasons.

Because of the need to ensure coverage of all contingencies, they can rarely be less than 2 hours long, and often have to be extended over a full working day.

(*b*) *Partially summarize the studies*

Initially, the following data is required to be extracted from the studies in order to use them as a check—

y = the output details. (1,080 dials 4 in. diameter; 240 dials 8 in. diameter; 86 dials 12 in. diameter).
e = the effective time of the study (269·21 minutes).
r = the average rating (106%).
t = study time (297·71 minute).
l = the waiting time (4·4 minutes).
L = any other ineffective time except relaxation (nil).

(*c*) *Check standard times*

Rating is the rate of producing basic time.
Performance is the rate of producing measured work units.

The difference between basic time and work units is relaxation allowance.

Therefore, an operator who works at a given *rating* and takes precisely the relaxation allowed for in the standard times will be working at the same *performance*.

To check standard times, therefore, the same allowance for relaxation must be made in the checking period as has been allowed in the times themselves when the performance can be compared with the average rating.

One way of doing this is to determine the operator performance which would have been achieved if the operator had produced the output during the study in a period equal to the net effective time plus the average relaxation percentage allowed for in the standard times, and to compare this with the average rating. If the two are within $\pm 2\frac{1}{2}$ per cent of each other, the standard times can be said to be valid since no further limit of accuracy can be guaranteed for measured work.

The procedure is explained as follows—

(i) Determine the total value of the measured work produced during study from—

$$W = v_1 y_1 + v_2 y_2, \text{ etc}$$

where $y =$ the output during the study

$v =$ the computed standard times (4 in. diameter, 19 standard minutes per 100; 8 in. diameter, 31 standard minutes per 100; 12 in. diameter, 50 standard minutes per 100).

In the *example*, this was—

$$W = (1080 \times 19) + (240 \times 31) + (86 \times 50)$$
$$= 20520 + 7440 + 4300$$
$$= 32260 \text{ standard centiminutes}$$
$$= 322 \cdot 6 \text{ standard minutes}$$

(ii) Determine the average relaxation allowance percentage which has been included in the computed standard times.

This is obtained by inspection of the synthetic or other data. In the *example* it was—

$$R = 14\%$$

(iii) Determine the theoretical time it would have taken to produce the output during the study, assuming there was no waiting time or other losses, and that exactly the amount of relaxation allowed R was in fact taken, i.e.—

$$H = \frac{(100 + R)e}{100}$$

In the *example* this was—

$$H = \frac{(100 + 14)269\cdot21}{100}$$

$$= 306\cdot9 \text{ minutes}$$

(iv) Determine the operator performance which would have resulted from working on the computed standard times in the gross time H from—

$$p = \frac{100W}{H}$$

In the *example* this was—

$$p = \frac{100 \times 322\cdot6}{306\cdot9}$$

$$= 105$$

(v) Compare the operator performance with the average rating. If it is within $\pm 2\frac{1}{2}$ per cent certify the standard times as valid. If a greater error exists, carry out the next stage (*d*).
In the *example* this was—

operator performance p = 105
the average rating r = 106

Since these were within less than $\pm 2\frac{1}{2}$ per cent of each other, the standard times were acceptable.

(vi) Determine the overall performance from

$$P = \frac{100W}{t - (I + L)}$$

In the *example*, this was—

$$P = \frac{100 \times 322\cdot6}{297\cdot71 - (4\cdot4 + \text{nil})}$$

$$= \frac{32260}{293\cdot3I}$$

$$= 110$$

which means that the performance actually made by the operator during the study exceeded the average rating. This was because the relaxation allowance taken was less than that allowed. If there are large discrepancies between this value and the operator performance p, the relaxation allowance standard should be checked (see step (*e*)).

(*d*) *Further summarize the studies and adjust standard times where necessary*

In those cases where inaccuracies in standard times are revealed after checking (i.e. that there is a greater discrepancy than $2\frac{1}{2}$ per cent between the operator performance and the average rating), it will be necessary to complete any studies used for checking up to the stage described in Chapter 3, and to compare the data element by element and frequency by frequency until the error has been discovered. Suitable adjustments are then made—or further studies taken if necessary—until completely satisfied that the synthetic or other data is accurate and soundly based.

It is most important that this check is carried out thoroughly before any standard times are released for issue.

(*e*) *In special cases check relaxation allowances*

Where the operator performance *p* and the overall performance *P* varies considerably, it may be necessary to carry out an independent check of the relaxation allowance actually taken. This is calculated from

$$RA\% = \frac{100\,(\text{idle time})}{e}$$

where $e =$ the effective time and the idle time is that part of the time where the operator had work available but did not do it.

For example, if the *RA* percentage as calculated by this means was 23 per cent, but the average relaxation allowance included in the standard times was 19 per cent and an observer taking a production study felt that all the relaxation taken during the study was in fact necessary, then some adjustments to these values would have to be made to bring the standard times into line.

8

Broad work measurement

The work measurement techniques so far described apply to operations which are mainly of a repetitive nature—or at least repetitive inasmuch as they consist of a series of identical and comparably similar elements of short duration joined together in different ways.

A great deal of work carried out in industry, however, does not fall into this category. Examples are found in maintenance departments, jobbing shops, tool rooms, offices and drawing offices, etc., where many of the operations take several hours or days to complete. This makes it largely impracticable to measure by conventional stop watch methods. But if it is analysed on a broader basis, the pattern revealed will be the same, i.e. that the work is composed of different combinations of identical and comparably similar elements, the only differences being that the element times will be longer and somewhat less consistent in accuracy.

Nevertheless, with such data available it is possible to estimate overall operation times closely enough to use them for incentive payment and other purposes. Work standards for use as synthetic data to build up work values can be obtained by using either one of the following techniques or by combining some of them together, the choice being determined by the character of the work and other circumstances—

1 Work measurement using an ordinary wrist watch or clock.
2 Work measurement by direct observation without the use of a time piece.
3 Work measurement from examination of work records.
4 Analytical estimating.

5 Activity sampling.
6 Analytical observation.
7 Work measurement using predetermined standard time data.

Work measurement using an ordinary wrist watch or clock
This method is very similar to that used for time study (reference Chapter 3), except that instead of a stop watch, an ordinary wrist watch or clock is used, preferably one fitted with a seconds-hand. Very often, more than one operator can be studied at a time, and a special study board designed to hold up to four observation forms at one time has been used successfully for work measurement in a maintenance shop and a tool room.

The procedure for taking these studies is as follows—

(*a*) Discuss the investigation with supervision and labour and select suitable operators for preliminary study.

(*b*) Analyse the operation into work elements, time and rate each, and record all other happenings.

(*c*) Have the quality of the work checked and record details of the operation and workplace.

(*d*) Thank the operators for their co-operation.

(*e*) Summarize study and complete top sheet.

Each step is now explained in more detail.

(*a*) *Discuss the investigation with supervision and labour and select suitable operators for preliminary study*
This preliminary approach must be made whenever work measurement is carried out and is described in more detail in Chapter 3.

(*b*) *Analyse the operation into work elements, time and rate each, and record all other happenings.*
Observations are recorded on *work study observation sheets* (Fig 9) clipped to a study board, up to four being used at one time, if possible, each being marked with details of the job to be studied. As each job commences, record the time from the wrist watch to the nearest half minute at the head of the appropriate sheet. Rating need not be so accurate as for time study and can be carried out to the nearest 10 every half minute or so. Division into work elements will be very broad. For example, turning, milling or shaping operations could be divided into elements as follows—

Procure tools.
Set workpiece in machine.
Machine one length and check dimensions.
Remove workpiece from machine, inspect and lay aside.
Put away tools.

The type of machine, essential dimensions of the part, specification of material, etc., should also be carefully recorded on the observation sheets.

Measurement of office work may include elements such as—

    Check goods received list with delivery promises.
    Answer telephone enquiry from sales department.
    Type shortage list.
    Record stores receipts on stock control cards.

When studying work such as painting and decorating, overall times can be taken for preparing and painting previously measured areas of wall and ceiling using different paints on various types of surface. Times could also be recorded on painting complete doors and windows of standard sizes as well as known lengths of skirting boards, etc.

As when carrying out time study, all ineffective times should be noted whenever they occur as well as the time at the conclusion of each study.

(*c*) *Have the quality of the work checked and record details of the operation and workplace*

It is important to verify the quality of any work which has been studied. This applies equally well to office work as to work done elsewhere. Time standards set on copy typing for example, will not be valid if the finished results contain so many errors that a considerable amount of extra time has to be spent in rectification.

(*d*) *Thank the operators for their co-operation*

This simple courtesy should never be forgotten, as it will tend to preserve the good relations so essential for effective work study.

(*e*) *Summarize study and complete top sheet*

Studies should be summarized on *work study observation and record sheets* (Fig 12). Since the continuous timing method is used, elapsed element times must be obtained by subtraction and normalized to basic times by applying the rating factor. Each element should be given a letter reference, these being used throughout summarizing, and finally entered to the *top sheet* so that elements can be quickly traced to their originating observation sheets.

The *work study top sheet* (Fig 13) should finally be completed with details of all the elements, basic elemental times, ineffective times, relaxation periods, elapsed and net effective times together with the average rating, finally stapling the sheets together as explained in Chapter 3.

Work measurement by direct observation without the use of a time-piece

Sometimes, operators will object to the use of any form of timing device. One of the ways of overcoming this is to use a combined technique of direct observation and estimation. The procedure is exactly the same as that described using a wrist watch or clock, except that only the overall times of the jobs are noted, the individual work elements themselves being rated and estimated during the study. Adjustment to these estimates may then have to be carried out on summarizing, by ensuring that the total elemental times is always equal to the net effective times of the study.

This method, although not so accurate as that previously described, will produce surprisingly satisfactory results when later comparison is made between similar work elements to obtain standards.

Work measurement from examination of work records

To use this method, operators are requested to fill in a time sheet for a limited period which covers the whole of the time in attendance. In doing this, they may also be required to record quantities of parts produced, and to break down the work into broad intervals of time. Examples from the records kept for painting and decorating in a maintenance department could be—

Washing down and stopping walls of office—4 hours.
Washing down and stopping ceiling of office—2 hours.
Sand woodwork—3 hours.
Apply first coat of emulsion paint to walls—$1\frac{1}{2}$ hours.

The forms used should be carefully designed to collect information broken down into broad statements that are not too tedious to record. Its intention and use should be personally explained to each operator or person concerned, the analyst actually completing some of the forms to give encouragement and ensure accurate records. In effect, the information should be similar to that which would be collected using direct observation methods, except that the statements may have to be broader in character to reduce detail to a minimum.

To summarize the information for use as time standards, some measure of a performance index must be obtained for each of the operators or for the department as a whole. A trained observer who is spending a great deal of time in the department should be able to make a reasonably close assessment of this, particularly if assistance is given by a co-operative foreman or overseer.

Normalizing to basic times is then carried out on *work study observation sheets*, using these assessed rating factors.

A *work study top sheet* is then completed in the same manner as that used for time study, except that the average basic time will generally be inclusive of relaxation allowance and will have to be marked as such on the study.

Work sheets are stapled to completed studies and the whole given a study number from the *study register*.

Analytical estimating

A skilled craftsman or foreman, by reason of his training and experience, can often estimate overall times for jobs of short duration with reasonable accuracy. Where the work is longer and more complex, however, this method is not so reliable unless it is broken down into smaller, more manageable portions.

If, therefore, an experienced tradesman is given basic training in work measurement techniques, with particular regard to the methods used to analyse work into separate elements, he should be capable of estimating operations of fairly long duration quite accurately.

A simple example to illustrate the method is given in Fig 27 which is an estimate prepared for painting and decorating an office. The work has been analysed by measurement of the various areas of wall, ceiling, number of doors and windows, condition of paintwork, etc., and each item of preparation and painting considered separately to build up the final estimated standard time.

Some work is difficult to estimate in this way. Examples include assessing times to repair equipment such as compressors, electric motors, special purpose machines, etc., that have broken down. In these instances the work content may vary as much as ± 70 per cent or more according to the faults found. One method of overcoming this problem is first to assess the average time for doing the job; then to decide how long it would be likely to take if the worst happened and it became a major operation (the most pessimistic estimate); and finally to give an estimate assuming the most favourable conditions (the most optimistic estimate). The average expected time can then be calculated from the formula—

$$E = \frac{c + 4l + p}{6}$$

Where E = the average expected time.

c = the most optimistic estimate (i.e. the least time expected).

p = the most pessimistic estimate (i.e. the longest time ex-expected).

l = the most likely estimate.

REGISTER OF CONSTANT ELEMENTS SHEET

Department __Maintenance__   Set up No. __Estimate No 76__

Section __Painters__   Date __19·8·71__

Product __Works Engineer's office__ Sheet No. __1__ of __1__

Operation __Paint and decorate__   Type of set-up __Estimate__

__complete.__   Plant details _____

_____

Remarks __Individual times in__   _____

__Standard minutes each. Final__   _____

__time in standard hours.__ *   _____

Ref. No.	Summary ref.	Element	Selected basic mins. per 100	% R.A.	% Cent	Standard minutes per 100 occs	Freq. basis	Std. mins. per 100
1		Procure paint and tools etc	per	occ		7.0	1	7
2		Move furniture and lay						
3		protective sheets	per	occ		5.0	1	5
4		Wash and sand walls.	sq.	yd		1.5	50	75
5		Wash and sand ceiling	sq.	yd		2.0	20	40
6		Rub down and stop doors.	per	door		15.0	2	30
7		Rub down and stop windows	per	window		20.0	2	40
8		Rub down & stop skirting board	per	foot		1.1	56	62
9		Emulsion paint walls	per sq. yd.			5.0	50	250
10		Emulsion paint ceiling	per sq. yd.			6.0	20	120
11		Paint doors with undercoat & top coat	per door			40.0	2	80
12		Paint windows with undercoat & top coat	per window			60.0	2	120
13		Paint skirting board with						
14		undercoat & top coat	per ft			2.0	56	112
15		Take up protective sheets						
16		and put back furniture	per	occ		5.0	1	5
17		Replace paint, tools etc						
18		into workshop and stores	per	occ		7.0	1	7
19								
20								
21								
22								
23								
24								
25								
				TOTALS				953

WORK VALUE IN STANDARD MINUTES PER (Standard Hours) | 16 *

COMPILED BY   S. F. H.

Fig 27 Register of constant elements sheet being used for the preparation of an analytical estimate

To quote an example, assume three estimates were made for repairing a small compressor as follows—

The most optimistic estimate $c =$ 1 hour
The most pessimistic estimate $p =$ 11 hours
The most likely estimate $l =$ 3 hours

Then according to this method, the average expected time will be—

$$E = \frac{1 + (4 \times 3) + 11}{6}$$

$$= 4 \text{ hours}$$

If jobs of this nature can be analysed into smaller parts and each estimated separately by this means, the total estimated time will tend to become more accurate. For example, if repairing the compressor consisted of—

1 Remove compressor from mounting.
2 Carry out repair.
3 Replace compressor on mounting.

and each operation is independently assessed, the three average expected times added together will give a more accurate result than if the job was estimated as a whole.

The method is used extensively in *PERT* (Program Evaluation Review Technique), which is a form of *network analysis* (see Chapter 19).

Activity sampling
When it becomes necessary to measure work of very long duration, or to investigate the activities of operators or machines as a group, it is often quite uneconomical to achieve this by using time study. If, however, a number of instantaneous observations are made at random intervals, the results can be used to discover the amounts of activity and non-activity and the proportions in which the various events are spread over the period observed. Also, where operators are being studied, and assessments are made of their rates of working in addition, the data can be used to develop standard times.

These methods are known as *activity sampling* and *rated activity sampling* respectively. They can be explained more clearly by considering a few examples of their use.

EXAMPLE 1
A single operator working over a 5-hour period on a simple job such as covering walls of a small office with emulsion paint, might distribute his activities as shown in Fig 28.

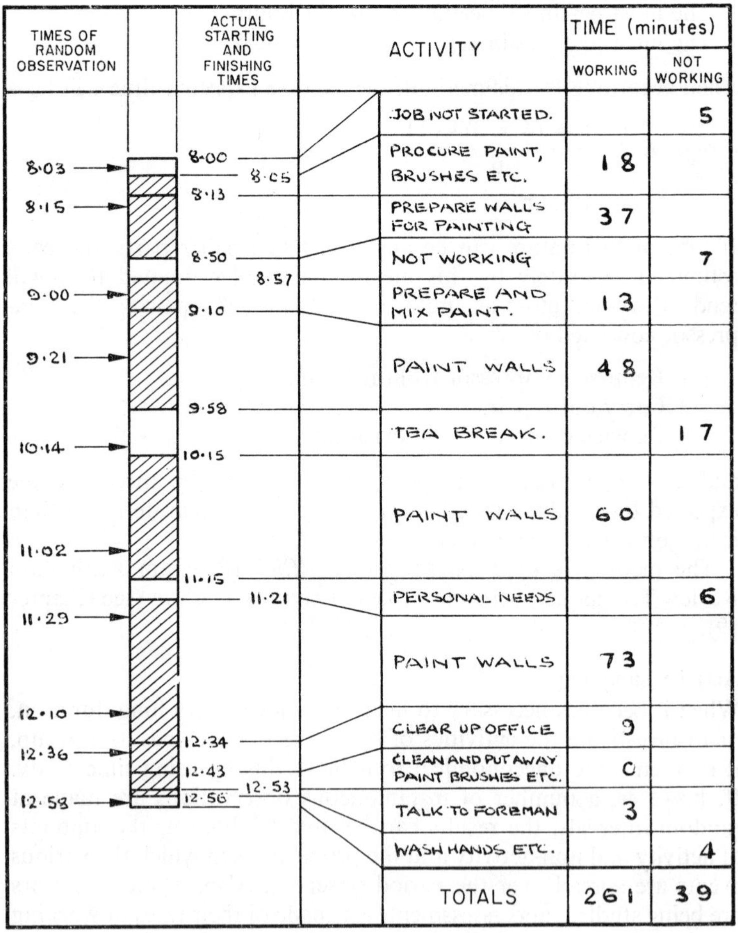

TIMES OF RANDOM OBSERVATION		ACTUAL STARTING AND FINISHING TIMES		ACTIVITY	TIME (minutes)	
					WORKING	NOT WORKING
				JOB NOT STARTED.		5
8·03 →		8·00 — 8·05		PROCURE PAINT, BRUSHES ETC.	18	
8·15 →		8·13		PREPARE WALLS FOR PAINTING	37	
		8·50 8·57		NOT WORKING		7
9·00 →		9·10		PREPARE AND MIX PAINT.	13	
9·21 →				PAINT WALLS	48	
		9·58				
10·14 →		10·15		TEA BREAK.		17
				PAINT WALLS	60	
11·02 →		11·15				
		11·21		PERSONAL NEEDS		6
11·29 →				PAINT WALLS	73	
12·10 →				CLEAR UP OFFICE	9	
12·36 →		12·34		CLEAN AND PUT AWAY PAINT BRUSHES ETC.	0	
		12·43 12·53		TALK TO FOREMAN	3	
12·58 →		12·56		WASH HANDS ETC.		4
				TOTALS	261	39

Fig 28 Diagrammatic representation of the information likely to be collected by random observation of an operation of painting the walls of a small office with emulsion

These are further summarized below—

Activity	Time (*minutes*)	Per cent of total
Procure and put away paint, brushes, etc.	18	6·0
Prepare walls for painting	37	12·3
Prepare and mix paint	13	4·3
Paint walls	181	60·4
Clear up office	9	3·0
Talk to foreman	3	1·0
Total net effective time	261	87·0
Ineffective time (not working)	39	13·0
Total elapsed time	300	100·0

If instantaneous observations were made at the times shown in Fig 28, and the results of these tabulated as follows—

Time	Activity	Working	Not working
8·03	Job not started	–	1
8·15	Prepare walls for painting	1	–
9·00	Prepare and mix paint	1	–
9·21	Paint walls	1	–
10·14	Tea break	–	1
11·02	Paint walls	1	–
11·29	Paint walls	1	–
12·10	Paint walls	1	–
12·36	Clear up office	1	–
12·58	Job finished	–	1
	Totals	7	3

103

Further summarizing these, and comparing them with the actual times gives—

Activity	Number of random observations	Per cent observations of total	Actual per cent of total	Per cent error on actual
Procure and put away paint, brushes, etc.	—	—	6·0	−6·0
Prepare walls for painting	1	10	12·3	−2·3
Prepare and mix paint	1	10	4·3	+5·7
Paint walls	4	40	60·4	−20·4
Clear up office	1	10	3·0	+7·0
Talk to foreman	—	—	1·0	−1·0
Total net effective time	7	70	87·0	−17·0
Ineffective time (not working)	3	30	13·0	+17·0
Total elapsed time	10	100	100·0	

If, however, the number of observations taken at random is increased to 30, results are as follows—

Activity	Number of random observations	Per cent observations of total	Actual per cent total	Per cent error on actual
Procure and put away paint, brushes, etc.	1	3·3	6·0	−2·7
Prepare walls for painting	4	13·4	12·3	+1·1
Prepare and mix paint	1	3·3	4·3	−1·0
Paint walls	20	66·7	60·4	+6·3
Clear up office	1	3·3	3·0	+0·3
Talk to foreman	—	—	1·0	−1·0
Total net effective time	27	90·0	87·0	+3·0
Ineffective time (not working)	3	10·0	13·0	−3·0
Total elapsed time	30	100·0	100·0	

This indicates that the greater the number of observations taken, the more accurate the results become.

If, during the study, the operator is rated over short intervals of time, results could be summarized as follows—

Ratings

70, 90, 100, 80, 60, 80, 70, 105, 95, 70, 60, 80, 80, 90, 80, 70, 60, 100, 90, 80, 80, 60, 90, 70, 80, 90, 70, 100, 80, 70 (30 observations)

The total of these is 2,400. The average rating is, therefore, assessed at—

$$\frac{2,400}{30} = 80 \text{ BSI}$$

This average rating can be used to convert assessed actual times to basic times, and by adding the appropriate relaxation allowances, to standard times per activity.

In the table overleaf, 30 observations were taken at random over an elapsed time of 300 minutes (i.e. between 8.00 am and 1.00 pm, as shown in Fig 28). The assessed times in column (b) were obtained by multiplying the percentages in column (a) by 300, and dividing by 100. The basic minutes in column (c) were obtained by extension of the values in column (b) at the average rating of 80 BSI. These basic minutes were finally converted to standard minutes per activity in column (e) by adding the percentage relaxation allowances in column (d).

If the areas of the walls are known, the painting and preparation times in column (e) can be expressed as standard times per unit area, and the time values used to form the basis for synthetic work measurement data.

Activity	(a) Per cent observations of 30 total	(b) Number of minutes based on (a)	(c) Basic minutes (80 BSI rating)	(d) Per cent RA	(e) Standard minutes
Procure and put away paint, brushes, etc.	3·3	10	8	15	9
Prepare walls for painting	13·4	40	32	$17\frac{1}{2}$	38
Prepare and mix paint	3·3	10	8	15	9
Paint walls	66·7	200	160	$17\frac{1}{2}$	188
Clear up office	3·3	10	8	$12\frac{1}{2}$	9
Total net effective time	90·0	270	—	—	—
Total elapsed time	100·0	300	—	—	—

Total basic minutes 216   —   —

Total standard minutes 253

EXAMPLE 2

Consider a group of 10 machines on which there are various stoppages due to setting, re-setting, inattention and maintenance faults.

Observations made at random might give the following results—

Activity	Number of observations	Per cent of total observations
Machines running	143	55
Stopped due to setting	41	16
Stopped due to inattention	57	22
Stopped due to lack of maintenance	19	7
Totals	260	100

By this means, useful information can be collected showing where major losses are occurring and indicating those areas which will repay more intensive study.

EXAMPLE 3

Consider a group of 5 office workers whose main activities can be broadly described as (see Fig 29, p. 109)—

Type orders
Type letters
File orders
File letters
Operate photocopying machine
'Phone
Fill and seal envelopes

Random observations taken by making periodic tours of the working locations of these operators, recording idle or ineffective time and the nature of any activity being performed together with an assessment of the rate of working could be summarized as follows—

Activity	Number of observations	Per cent of total observations
Type orders	201	17
Type letters	263	23
File orders	47	4
File letters	71	6
Operate photocopying machine	57	5
'Phone	293	25
Fill and seal envelopes	84	7
Not working	153	13
Totals	1169	100

If the total time over which the study was taken was 150 hours, and the average of all the ratings 70 BSI and quantities of orders and letters filed and typed, the number of photocopies taken, and the number of envelopes sealed was also recorded, then further results could be computed as follows—

Activity	(a) Per cent of total observations	(b) Estimated time taken (minutes)	(c) Basic time (minutes) 70 per cent (b)	(d) Output
Type orders	17	1530	1071	83 orders
Type letters	23	2070	1449	74 letters
File orders	4	360	252	87 orders
File letters	6	540	378	71 letters
Operate photo-copying machine	5	450	315	153 documents
'Phone	25	2250	1575	—
Fill and seal envelopes	7	630	441	78 envelopes
Not working	13	1170	—	—
Total time (minutes) = (= 150 hours × 60)		9000		

Approximate standard times can then be obtained by adding appropriate relaxation allowances to the basic time and dividing these by the output.

Accuracy and number of observations for activity sampling

It can be proved statistically that the percentage accuracy over the whole activity can be predicted within 95 per cent confidence limits by the use of the following formula.

$$L = 2\sqrt{\frac{p(100 - p)}{N}}$$

Where L = the accuracy of the finished result expressed as a percentage of the total time.

p = the proportion of any activity (or non-activity) expressed as a percentage of the total time.

N = the number of instantaneous observations made at random intervals.

and conversely—

$$N = \frac{4p(100 - p)}{L^2}$$

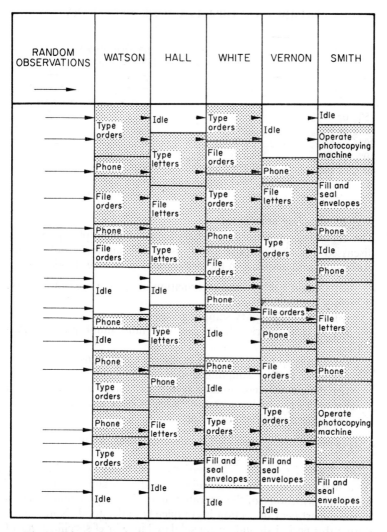

Fig 29 Diagrammatic representation of random observations being made on a graph of office workers by periodic tours

The following table will give some idea of the practical figures involved—

NUMBER OF OBSERVATIONS NECESSARY

	Accuracy of finished results (95% confidence limits)		
p	+ or − 5%	+ or − 10%	+ or − 15%
5% or 95%	76	19	9
10% or 90%	144	36	16
20% or 80%	256	64	29
30% or 70%	336	84	38
40% or 60%	384	96	43
50%	400	100	45

Activity sampling procedures

The following procedures are based on the principles previously discussed for the purpose of investigating machine and labour utilization, and assisting in the determination of time standards—

(a) Obtain background information about the area of activity.

(b) Plan study programme.

(c) Take preliminary sampling studies.

(d) Decide accuracy limits and determine number of observations.

(e) Draw up programme of random observations.

(f) Take further sampling studies.

(g) Periodically consolidate results of studies and cease when satisfactory results have been obtained. Check accuracy.

(h) Determine final utilization values.

This procedure is explained in more detail as follows—

(a) Obtain background information about the area of activity

As a first step, discussions should be held with supervision and, if necessary, with some of the operators until the purpose and function of the department or section is completely understood. Notes taken during this period should be consolidated into brief specifications of each of the machines or work of the operators about to be studied.

(b) Plan study programme

This entails determining the locations which are to be visited during the study and preparing *observation forms* for this purpose.

For example, if it is intended to investigate the proportion of running time to idle time on a number of machines, a *work study observation and summary sheet* could be made out as shown in Fig 30, the numbers (or other indentifications) of the machines being entered at the head of the columns, and running or not running times being indicated by a tick or a cross respectively.

Where operators are being studied, it may be necessary to consider various possible working locations and arrange the study programme around these. For instance, 5 office workers might carry out their activities in any of the following places—

Name			*Work stations*
Watson		A	(desk)
Smith		B	(desk)
Vernon		C	(desk)
	or	D	(filing cabinet)
White		E	(desk)
	or	D	(filing cabinet)
Hall		F	(desk)
	or	G	(photo-copying machine)

If *work study observation and summary sheets* are prepared as shown in Fig 30, and rounds made of points A to G, every one of the five operators should be located if they are legitimately at work. The activities themselves can be written in the extreme left-hand column, a cross being entered if the operator is not working or absent and a tick or rating evaluation if working.

(c) Take preliminary sampling studies

At this point, a few rounds should be made at fairly random intervals and observations taken. It is most important that these are instantaneous, i.e. if an operator is not working at first glance but starts immediately afterwards, this should still be recorded as "not working". Ratings should be assessed over very short intervals, otherwise there may be a bias due to the presence of the observer being noticed.

(d) Decide accuracy limits and determine number of observations

More will have been learnt about the operation from the preliminary studies. This may mean adjustments having to be made to the study programme, if, for example, extra legitimate work stations are discovered. From then on the procedure is as follows—

111

WORK STUDY OBSERVATION AND RECORD SHEET

Product **General office**   Study No. **756**

Operation **Office activity sampling**   Sheet No. **2** of **5**

Operators **Watson, Hall, White, Vernon and Smith**   Date **17.8.71**

Location	A		B		C		D		E		F		G	
	80		x		..		70		50		-		80	
	90		90		-		70		60		-		x	
Type orders.														
	70		60		-		70		60		..		70	
Type letters														
	80		90		-		60		70		-		x	
File orders.														
	-		x		80		90		70		60		-	
File letters														
	-		x		60		-		x		70		80	
Operate photo-copying machine														
	x		60		-		70		80		100		100	
Phone														
	100		90		-		60		80		x		-	
Fill and seal envelopes.														

Fig 30 Work study observation and record sheet shown partially complete and in use for rated activity sampling on a group of office workers

(i) Assess the percentage occurrence of the activities to the non-activities from the following formula—

$$P = \frac{100q}{Q}$$

where p = number of observations of activities as a percentage of total observations.

q = number of observations on all the activities separately.

Q = total number of observations taken (on both activities and non-activities).

For example, if a total of 10 observations have been made, 7 of which recorded a machine running or an operator working, and 3 of which recorded a machine stopped or an operator not working, then—

$$P = \frac{100 \times 7}{10}$$

$$= 70\%$$

or, in other words, the assessed percentage of activity to the total time is 70 per cent.

(ii) Decide on accuracy required and assess number of observations. This is obtained from the formula—

$$N = \frac{4p(100 - p)}{L^2}$$

For example, assuming an accuracy of ± 5 per cent is sufficient and

$$p = 70\%$$

$$N = \frac{4 \times 70(100 - 70)}{5^2}$$

$$= 336 \text{ observations}$$

(e) Draw up programme of random observations
This stage is carried out as follows—

(i) Decide the period of time over which the study is to be taken. This will be assessed from knowledge of the work.

For example, if a general office is to be investigated, and the routines tend to occur over a weekly cycle, this may decide the duration of the study. Other factors to be taken into consideration are the amount of work study observer's time available, which may restrict the number of observations possible in any given period.

(ii) Draw up a time schedule of random observations.

This can be carried out using tables of random numbers, and selecting only those which are between 1 and 59. For example, if the study is to be taken on the first day from 9 am until 1 pm and from 2 pm to 6 pm, and the first numbers to be selected on this basis are 36, 51, 24, 15, 3, 14, 47, 9, 31, 45, 17, then these could be written down as 9·36, 10·51, 11·24, 12·15, 2·03, 3·14, 4·47, 5·09 then 9·31, 10·45, 11·17, etc., until the required number of observations have been obtained. Another method is to prepare two sets of cards, one set numbered 0 to 59, and the other 1 to 12. By selecting cards in the smaller pile to represent the hours over which the study is to be taken (in the example cards 2, 3, 4, 5, 9, 10, 11 and 12 only would be used); shuffling the two packs separately, then drawing a card from each pack representing the hour and the minutes past the hour respectively. Replacing, re-shuffling and re-drawing cards will eventually enable a list of random times to be compiled.

These can then be set down on a prepared set of *work study observation and summary sheets* (see Fig 30).

(*f*) *Take further sampling studies*

During this period, it is of advantage to use more than one observer taking studies in order to relieve the monotony. Statements of activity should in general be broad, and confined to as few as possible. For example, in a general office the activities might be described as in Example 3, i.e. type orders, type letters, file orders, file letters, operate photocopying machine, 'phone, fill and seal envelopes, etc.

It is emphasized that the observations must at all times be instantaneous.

(*g*) *Periodically consolidate results of studies and cease when satisfactory results are obtained*

To ensure that no more observations are taken than are absolutely necessary, graph the results each day (or other convenient period) as shown in Fig 31, for each activity, plotting the total number of observations against the percentage of the total number taken.

For example, if one of the activities is "type letters" in a general office application and the recorded details so far are as follows—

Number of observations recorded as "type letters" = 25

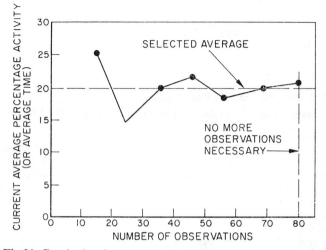

Fig 31 Graph showing percentages of total activity (or current average times where rated activity sampling is used), plotted against the number of observations

Total number of observations so far recorded on
$$\text{all operations} = 125$$

then "type letters" observations as a percentage of the whole

$$= \frac{25 \times 100}{125}$$
$$= 20\%$$

These values should be plotted against each other until the graph appears to have levelled out sufficiently when a selection can be made of the probable result. Further study work should then cease on that particular activity.

The probable accuracy should finally be checked from—

$$L = 2\sqrt{\frac{P(100 - p)}{N}}$$

For example if

$P = 72\%$ (percentage activity over total elapsed time)

and

$N = 64$ (the number of observations taken on a particular activity)

115

Broad work measurement

then

$$L = 2\sqrt{\frac{72(100 - 72)}{64}}$$
$$= \pm 11\cdot 2\%$$

(h) Determine final utilization values
Final results should be tabulated in a form similar to the following—

Details	Number of observations	Approximate per cent of total	Approximate per cent accuracy
Activity 1	175	42·2	±7·4
Activity 2	96	23·2	±8·6
Activity 3	35	8·5	±9·6
Activity 4	24	5·8	±9·6
Activity 5	23	5·6	±9·9
Sub-total	353	85·3	±3·8
Delay 1	36	8·7	±9·6
Delay 2	25	6·0	±9·5
Sub-total	61	14·7	±9·2
Totals	414	100·0	—

Rated activity sampling procedure

If carrying out rated activity sampling, the same methods are used as previously described except that rating assessments will be made and output details recorded on the studies. From then on the procedure is as follows—

(*j*) Determine average rating of operators under observation.
(*k*) Determine basic time values.
(*l*) Determine standard time values.

These steps are now explained more specifically—

(j) Determine average rating of operators under observation
This can be obtained by taking the arithmetic mean of the rating assessments made. For example, if values were recorded as follows—

70, 90, 100, 80, 95, 105, 60, 70, 60, 90.

116

then the average assessed rating would be—

$$= \frac{820}{10}$$

$$= 82$$

Basic times will be more accurate if rating assessments are made on operators individually rather than on a group as a whole.

(*k*) *Determine basic time values*
Basic times are determined as follows—

    (i) Multiply the percentage of each activity studied by the elapsed time of the study.

    (ii) Adjust these values by applying the appropriate rating factor to each.

    (iii) Divide each value by the recorded output during the study to obtain basic times per occurrence.

Summarizing of the information should be carried out on a *work study observation summary sheet* (Fig 9) and finally entered to a *work study top sheet* (Fig 13), as outlined in Chapter 3.

(*l*) *Determine standard time values*
Standard times are determined by adding relaxation allowances to basic times per occurrence obtained under (*k*) above.

Analytical observation
This is a sampling technique which can be used by observers who either have a fair knowledge of the nature of the work being studied themselves, or who can obtain assistance from those who have. The method of collecting data is so simple to learn, however, that it can be taught to any intelligent tradesman within a few days. It is by far the quickest, and in many ways, the most reliable broad work measurement technique which can be used for studying non-repetitive work. It can also be adapted for repetitive operations, and is particularly useful for observing team work.

The method of carrying out this technique is as follows—

(*a*) View the site of the intended job before work takes place wherever possible and roughly draft out the proposed job method. Obtain the assistance of the foreman or other specialist whenever necessary.

(*b*) Using an ordinary wrist watch or wall clock, and a series of *work study observation forms* clipped to a study board, pay random visits to the activity centre commencing with the scheduled starting time of the job or thereabouts. At each visit note the exact time, the performance rating of the operator the instant that he is observed; and the condition and progress of the work in some fair detail.

(*c*) Make logical assumptions from these observations based on knowledge of the type of work, the method, the general progress, and the rate of working. Adjust the frequency of visits as the job progresses in order to collect as continuous a record as possible consistent with the type of study, and continue observations until the job is finished. Recruit assistance of the foreman if in doubt at any stage.

(*d*) After conclusion of the study, analyse the observations and assumptions to obtain actual times, obtain the average performance, and extend values to basic times.

(*e*) Enter the basic elemental times and other details to a *work study top sheet* (Fig 28).

The procedure can be better understood by quoting an example of its use as applied to the operation of painting a small office shown in Fig 28. The *observation sheet* for this was completed as shown opposite.

Time	Conditions found	Rating	Assumptions
8.03 am	Operator not working, no preparations made	—	Job not started
8.15 am	Operator preparing walls for paint. Only small area of wall completed. Paint and brushes on site.	70	Allowing 5 minutes to procure paint and brushes and approx. 2 minutes' work on wall, operator started work at 8.08 am.
9.00 am	Operator starting to mix paint, walls completely prepared for paint	90	Approx. 5 minutes of paint mixing operation completed. Walls finished at 8.55 am.
9.21 am	Walls being painted. Approx. half area of one wall painted	70	At rate of working and area completed, probably commenced to paint 15 mins. ago at 9.06 am.
10.14 am	Tea break. One wall now completely painted	—	Operator probably ceased work at 10.00 am (tea break time). Will probably recommence at 10.15 am.
11.02 am	Walls being painted. Painting now approx. half completed	80	Judging by progress and rate of working, operator probably worked continuously after tea break
11.29 am	Walls still being painted. Progress slower than at last visit	90	Operator either stopped work since last visit, or slowed up. Estimate 10 minutes stopped
12.10 pm	Walls now almost completely painted	80	Progress about average. No evidence of slowing up or stopping since last visit
12.36 pm	Operator just commenced to clear-up office. Painting complete	90	Walls probably completed at 12.35 am. Clearing up probably completed in 10 mins. at 12.45 am.
12.58 pm	Operator not working. All brushes cleaned and put away	—	Job finished. Allowing 10 mins. to clean brushes probably finished complete job at 12.55 am.

119

Summarizing this information, using a *work study observation sheet* for the purpose gives—

$$\text{Average rating} = \frac{70 + 90 + 70 + 80 + 90 + 80 + 90}{7}$$

$$= 81 \text{ BSI}$$

A table of assumptions is then drawn up, using a *top sheet*—

ACTIVITY	ASSUMPTIONS			
	Start	*Finish*	*Actual time (minutes)*	*Basic time at average rating (minutes)*
Procure paint, brushes, etc.	8.08	8.13	15	12·1
Clean and put away brushes	12.45	12.55		
Prepare walls for paint	8.13	8.55	42	34·0
Mix paint	8.55	9.06	11	8·9
Paint walls	9.06	10.00	184	149·0
	10.15 (less 10 minutes I.T.)	12.35		
Clear up office	12.35	12.45	10	8·1
Totals			262	212·1
Total elapsed time	8.00	1.00	300	
Ineffective time (by difference)			38	

Comparing this with actual conditions obtained by production study gave the following results—

ACTIVITY	VALUES OBTAINED BY PRODUCTION STUDY			*Basic minutes by AO study*	*Error*
	Actual minutes	*Rating*	*Basic minutes*		
Procure paint, brushes, etc., and put away brushes	18	75	13·5	12·1	−10·4%
Prepare walls for paint	37	80	29·6	34·0	+14·8%
Mix paint	13	70	9·1	8·9	−2·2%
Paint walls	181	80	145·0	149·0	+2·8%
Clear up office	9	90	8·1	8·1	Nil
Talk to foreman	3	80	2·4	—	−100%
Total			207·7	212·1	+2·1%

Several points emerge from this—

(*a*) Although only 9 observations were made, the total basic times are surprisingly close.

(*b*) The main operation of painting the walls is also very close.

(*c*) The inaccuracies seem to be mainly in those operations which are relatively less important by reason of their size. (Inaccuracies in work forming a small percentage of the whole have less effect on the overall inaccuracy than those which form a larger part.)

(*d*) If the number of observations is increased (which is what should happen in practice), the accuracy will improve.

(*e*) If the values obtained from one study are compared on a common basis with similar values obtained from other studies, inaccuracies can further be smoothed out.

(*f*) If then the finally selected values are in the form of synthetic data, any inaccuracies still existent in these will further be reduced when they are added together to form standard times per operation, by reason of the theory explained in Chapter 11.

(*g*) Analytical observation studies are ideal as production studies used to check standard times obtained for synthetic data (see Chapter 7).

Also, the technique need not be confined to the study of one operation alone. Up to 6 or more separate operations can easily be accommodated by one observer using this technique. The number can be increased even further if a portable tape recorder is used to record the observations at the activity centre, and then played back in the office to extract the essential information.

A *work study top sheet* for the examples shown completed in Fig 32, the sizes of the walls having been determined by measurement as 35 square yards, and the basic times for the preparation and painting elements expressed in basic times per square yard.

Element times can then be entered to study summary sheets and developed to synthetic data using the procedures explained in Chapters 4 to 7.

Predetermined motion-time systems

As is explained in Chapter 13, broad work measurement can be carried out by using one of the second generation predetermined motion time systems such as *MTM*–2, *SPMTS*, etc.

Practical applications

The choice of which technique to use for a particular application will be influenced by several factors, two of the most important being—

(*a*) The industrial relations climate.
(*b*) The nature of the work.

It should never be assumed that one overall technique is the best. Generally, it is probably better to use several techniques together rather than be confined to one above, providing that the labour will accept these conditions. A few suggestions of such combinations are given below—

(*a*) Use analytical observation for the main collection of data. Augment this with wrist-watch studies of small operations or parts of operations. For example an operator could be directly studied painting a known area of wall to check the basic time per unit area obtained by other means.

(*b*) Use direct observation without the use of a time piece combined with a system of *PMTS* where objection is raised to timing devices of any sort. If a tape recorder is permitted, rough times can be obtained from it, if the work is described into it as it takes place and the time subsequently obtained by relating this to the speed of the tape.

WORK STUDY TOP SHEET

Department __Maintenance__	Study No. __1001__
Section __Painters__	Date __7-8-71__
Product __Foreman's office__	Taken by __H.P.B.__
Operation __Paint walls__	Operator __J. Lawson__
__with white emulsion paint__	From __8.00a.m.__ to __1.00 p.m__
	Estimated Av. Rating __81__
No. of cycles _____ Output _____	Elapsed time __300 mins__
Plant details _____	Check time __—__
Remarks __Wall measurements__	Ineffective time __38 mins__
__35 square yards super.__	Net effective time __262 mins__

Element reference	Element	Basic time centimins.	Frequency	
A	Procure and put away paint, brushes & other tools	1210	1	per occurrence
B	Prepare plaster walls for paint with glass paper	97	35.	per square yard
C	Mix paint	890	1	per occurrence
D	Paint plaster walls with white emulsion paint	426	35	per square yard
E.	Clear up in office after completion of job	810	1	per occurrence

Fig 32 Work study sheet shown completed for analytical observation study given in Fig 28

(c) Use analytical estimating combined with stop-watch or wrist-watch studies. Where this is done, however, the elements obtained by estimating must be reduced by the proportion of their relaxation and contingency allowances to allow them to be compared on a common basis after entry on to *study summary sheets* as explained in Chapters 4 to 7. This also applies where examination of work records is used in conjunction with any of the direct or indirect work measurement techniques.

9

The measurement of team work

Unoccupied time and process allowance

Given a reasonable environment, adequate facilities, and an un-broken supply of work, the time taken by a single operator to complete a specific task unassisted by machine, will depend on the amount of skill and effort he exerts. When semi-automatic or automatic machines are used, or where the work is performed by a team or gang of operators, however, there may be a break in the continuity due to certain operators having to wait until the machines or other team members have completed their respective work cycles. This delay period is known as *unoccupied time*, and when in the form of an allowance made to operators working on bonus incentive, it is referred to as *unoccupied time allowance* or *process allowance*.

A simple incidence of this can occur in domestic life when two people are washing up. If the washing operation has the greater work content, and each is roughly working at the same pace, the person drying will be obliged to wait occasionally until there is sufficient work to keep him occupied, the total of his enforced idleness being the difference between the measured work content of the two tasks.

Multiple activity charts

Since unoccupied time is a form of waste, it must be reduced to the lowest possible proportions. In the example quoted, this can be brought about either by distributing the work differently, or by

125

doing other work in the idle period, such as putting away the dishes, etc. But on more complex operations the immediate solution is not so apparent, and some form of charted presentation is necessary to enable the facts to be seen more clearly. One of the methods of doing this is to represent each work or process cycle by a bar drawn to a time scale in the form of a diagram. Such diagrams are known as *bar charts* or *multiple activity charts*.

One of the uses of these can be briefly explained by quoting another example from domestic life—that of making tea. The procedure and standard times for this could be as follows (the times are not necessarily accurate)—

Work element	Standard time (*minutes*)
Procure and fill kettle with water and place on stove	0·5
Boil water in kettle	5·0
Procure teapot and tea	0·5
Procure and lay out tea cups, etc.	2·0
Pre-heat teapot with boiling water and load tea to pot	0·5
Pour boiling water from kettle into teapot	0·5
Allow tea to brew	1·0
Pour out milk, load sugar to cups	0·5
Pour out tea	0·5
Total	11·0 minutes

On the face of it, it appears that it would take 11 minutes to complete this job. If, however, a little care is taken in planning, it can be shown diagrammatically that it can be reduced to 8 minutes. if the work is performed in the following way. The total unoccupied time is 3 minutes (Fig 33)—

1 Procure and fill kettle with water and place on stove.
2 Procure teapot and tea.
3 Procure and lay out tea cups, etc.
4 Pour out milk, load sugar to cups.
5 Wait for kettle to boil (unoccupied time)
6 Pre-heat teapot with boiling water and load tea to pot.
7 Pour boiling water from kettle into tea pot.
8 Allow tea to brew (unoccupied time)
9 Pour out tea.

126

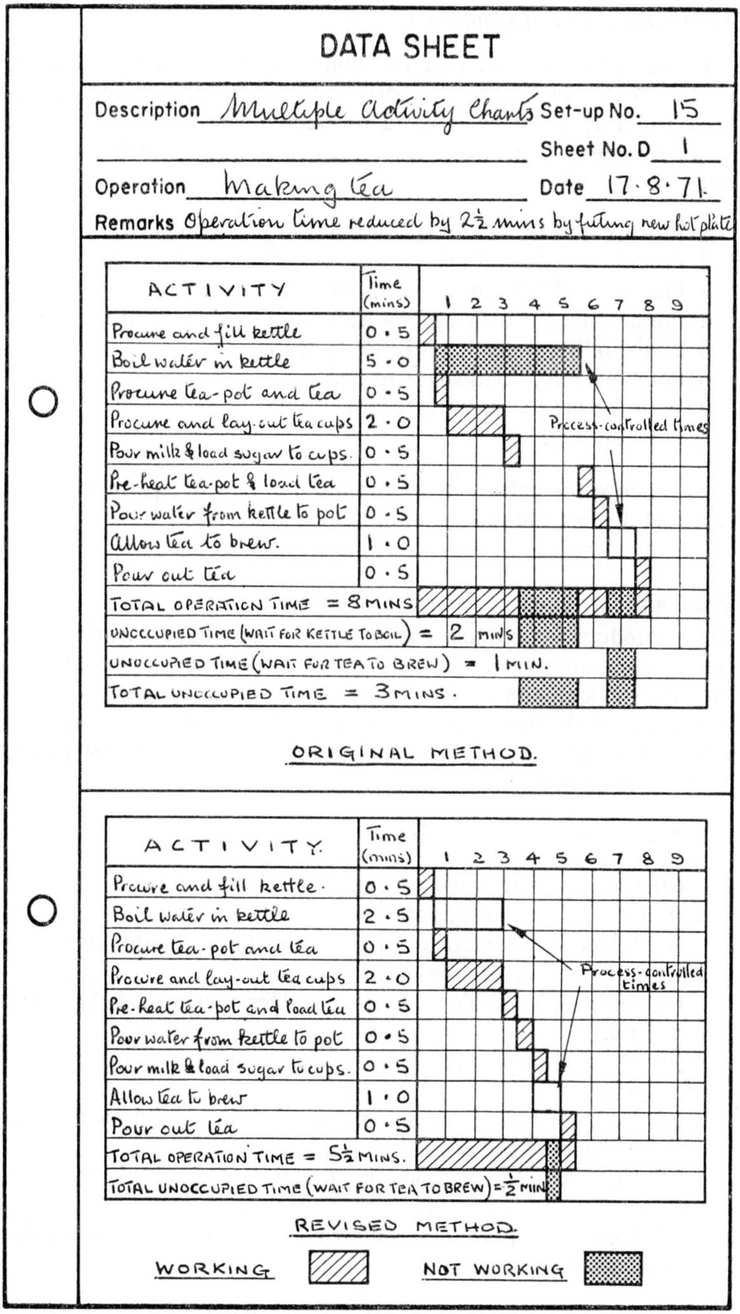

DATA SHEET

Description Multiple Activity Charts Set-up No. 15

Sheet No. D 1

Operation Making Tea Date 17·8·71·

Remarks Operation time reduced by 2½ mins by fitting new hot plate

ACTIVITY	Time (mins)	1	2	3	4	5	6	7	8	9
Procure and fill kettle	0·5									
Boil water in kettle	5·0									
Procure tea-pot and tea	0·5									
Procure and lay-out tea cups	2·0					Process-controlled times				
Pour milk & load sugar to cups.	0·5									
Pre-heat tea-pot & load tea	0·5									
Pour water from kettle to pot	0·5									
Allow tea to brew.	1·0									
Pour out tea	0·5									
TOTAL OPERATION TIME = 8 MINS										
UNOCCUPIED TIME (WAIT FOR KETTLE TO BOIL) =	2	mins							
UNOCCUPIED TIME (WAIT FOR TEA TO BREW) =	1 MIN.									
TOTAL UNOCCUPIED TIME = 3 MINS.										

ORIGINAL METHOD.

ACTIVITY	Time (mins)	1	2	3	4	5	6	7	8	9
Procure and fill kettle.	0·5									
Boil water in kettle	2·5									
Procure tea-pot and tea	0·5									
Procure and lay-out tea cups	2·0				Process-controlled times					
Pre-heat tea-pot and load tea	0·5									
Pour water from kettle to pot	0·5									
Pour milk & load sugar to cups.	0·5									
Allow tea to brew	1·0									
Pour out tea	0·5									
TOTAL OPERATION TIME = 5½ MINS.										
TOTAL UNOCCUPIED TIME (WAIT FOR TEA TO BREW) = ½ MIN										

REVISED METHOD.

WORKING ▨   NOT WORKING ▨

Fig 33 Multiple activity chart for "making tea"

If it is now assumed that the boiling cycle could be shortened to $2\frac{1}{2}$ minutes by fitting a special hot plate to the stove, the unoccupied time could be reduced to $\frac{1}{2}$ minute and the overall time to $5\frac{1}{2}$ minutes, if the work is performed in the following manner (see also Fig 33)—

1 Procure and fill kettle with water and place on stove.
2 Procure teapot and tea.
3 Procure and lay out tea cups, etc.
4 Pre-heat teapot with boiling water and load tea to pot.
5 Pour boiling water from kettle to teapot.
6 Pour out milk, load sugar to cups.
7 Wait for tea to brew (unoccupied time).
8 Pour out tea.

In practice the extra cost of running the new hot plate will have to be set against the reduction in labour cost.

Procedure for the measurement of team work
The procedure for initially collecting and collating information for time value computation on work which is performed by a team of operators is similar to that already described in Chapters 3 to 7 if using fine work measurement, or those described in Chapter 8 if using broad work measurement, except that certain extra steps will have to be carried out.

It can be summarized as follows—

(*a*) Collect basic work measurement data using time study or any of the other methods so far described.
(*b*) Select basic element times, compute and add relaxation allowances and determine basic work values.
(*c*) Determine individual time values performed by team members.
(*d*) Draw multiple activity charts to represent existing methods of working.
(*e*) Reduce or eliminate any unoccupied time and draw multiple activity charts of the revised methods.
(*f*) Test revised methods in practice, and check the time values.
(*g*) Determine final time values incorporating any unoccupied time allowances.

These steps are now considered in more detail.

(a) Collect basic work measurement data using time study or any of the other methods so far described

This consists of following the procedures outlined previously to collect basic work measurement data. Either time study or pre-determined motion-time data can be used for fine work measurement, adopting the routines described in Chapters 3 or 7 respectively. Broad work measurements are described in Chapter 8.

Careful note will have to be made of all those points when un-occupied time occurs, and these intervals recorded on the studies. Note must also be taken of the points when other operators cease their work relative to any operator who is being studied.

(b) Select basic elemental times, compute and add relaxation allowances and determine basic time values

Adopt the routines described in Chapters 4, 5 and 6 if using fine work measurement, or those described in Chapter 8 for broad work measurement.

(c) Determine individual time values performed by team members

This follows the procedure set out in Chapters 6 and 7. Some operations may have to be broken into intervals between unoccupied time occurrences.

(d) Draw multiple activity charts to represent existing methods of working

This consists of constructing a chart similar to that shown in Fig 33. The time values are represented by bars to a time scale.

(e) Reduce or eliminate any unoccupied time and draw multiple activity charts of the revised methods

Unoccupied time principally occurs in team working due to uneven distribution of duties between individual operators, or by the work not being performed in the best sequential pattern. It can often be reduced by applying one or more of the following remedies—

(i) Rearrangement of the sequence of work.
(ii) Re-allocation of duties.
(iii) Reduction of the governing time (or longest operation) by method study.
(iv) Allocation of extra work to fill the unoccupied time periods.
(v) Arranging conditions to allow operators to move between work stations, clearing the work as they do so.

Applications of these remedies are described in examples given at the end of this chapter.

(f) Test revised methods in practice, and check the time values
At this point, the work should be rearranged according to the revised methods, and tested on a trial basis. Time studies should be taken in order to check the new time values. Checking of these in this way is described in chapter 7.

(g) Determine final time values incorporating any unoccupied time allowances
Unoccupied time should not form part of the work value as such but should be published separately as an expendable allowance which can be withdrawn if working conditions change.

The following calculations are useful in obtaining these allowances and time values—

(i) The net time value.
This is the sum of all the net times necessary to complete the work, exclusive of unoccupied time.

If N = net time value
and p = operation times,
then $N = \Sigma p$

(ii) The governing time.
This is the longest operation or task being performed by a team of operators.

(iii) The gross time value.
This is the total time value inclusive of unoccupied time.

if G = gross work value
         l = the governing operation
and n = the number of operations being performed
then $G = nl$

(iv) The total unoccupied time per unit number of pieces.

if UT = unoccupied time per piece
then $UT = G - N$

(v) The unoccupied time per unit number of pieces per operation.
This may be different for each operation and is calculated as follows—

$$ut = l - p$$

where ut = unoccupied time per operation per unit number of pieces
          l = the governing operation time per unit number of pieces.
           = individual p operational time per unit number of pieces.

(vi) The percentage unoccupied time.
This is calculated as follows—

$$UT\% = \frac{UT}{N} \times 100$$

$$ut\% = \frac{ut}{p} \times 100$$

(vii) The operation efficiency can be calculated from—

$$E = \frac{N}{G} \times 100$$

Example 1—Reduction of unoccupied time on a team of two workers
A multiple activity chart for a team of two workers engaged on a simple operation is shown at A, Fig 34. The second operator is obliged to wait at periods, a, b, c and d because of lack of balance between the two operations.

The operations are presented more conveniently at B, showing the unoccupied and working times of the second operator as two separate blocks.

Reduction of both unoccupied time and overall cycle time was brought about by either—

(*a*) Re-allocation of work load, by transference of some of the work from the first to the second operator (Fig 34, diagram C).
(*b*) Introduction of a third operator to share the first operation (Fig 34, diagram D).
(*c*) Reduction of the first operation by method study (Fig 34, diagram E).
(*d*) Inclusion of extra useful work for the second operator (Fig 34, diagram F).

Example 2—Reduction of unoccupied time on a "flow line" team
This example refers to a small diecast component requiring several machining operations to be performed on it. The procedure used was as follows—

(*a*) Time values of each machining operation were obtained from synthetic data compiled by time study, using the procedures described in Chapters 3 to 7. Details of the operations were as follows—

131

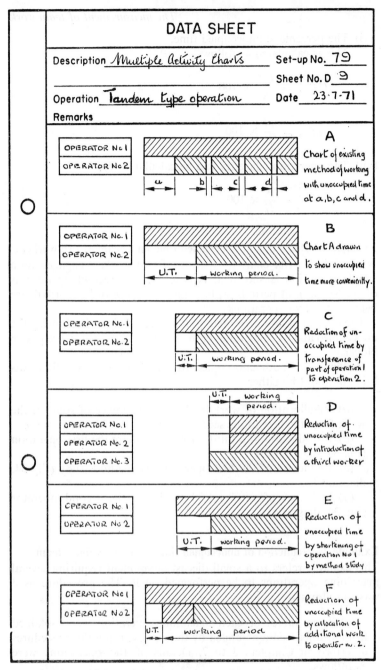

Fig 34 Multiple activity charts for a team of two workers and illustrating different methods of reducing unoccupied time

Operation	Time values (standard hours per 1,000 pieces)
1 Press from sprue	5·0
2 Drill 4 holes	6·0
3 Mill face	10·0 (governing operation)
4 Tap 4 holes	5·0
5 Linish face	3·0
6 Linish ends and sides	7·0

(b) A *multiple activity chart* was drawn for these values as shown at A, Fig 35.

(c) The net time value N was calculated by adding together the work values of all the operations.

$N = \Sigma p$

$\quad = 5 + 6 + 10 + 5 + 3 + 7$

$\quad = 36$ standard hours per 1,000 pieces

(d) The gross time value inclusive of unoccupied time G was next calculated. This is equal to the longest (or governing) operation multiplied by the total number of operations.

$G = nl$

$\quad = 6 \times 10$

$\quad = 60$ standard hours per 1,000 pieces

(e) The total unoccupied time was obtained by subtraction of the net time value from the gross time value.

$UT = G - N$

$\quad = 60 - 36$

$\quad = 24$ standard hours per 1,000 pieces

(f) Unoccupied time was reduced by including an extra operator to carry out operation number 3 (mill face). The effect of doing this is shown in the multiple activity chart at B in Fig 35.

The number of operations has now increased to 7, and the governing time is now operation number 6 (linish ends and sides). This is 7 standard hours per 1,000 pieces.

133

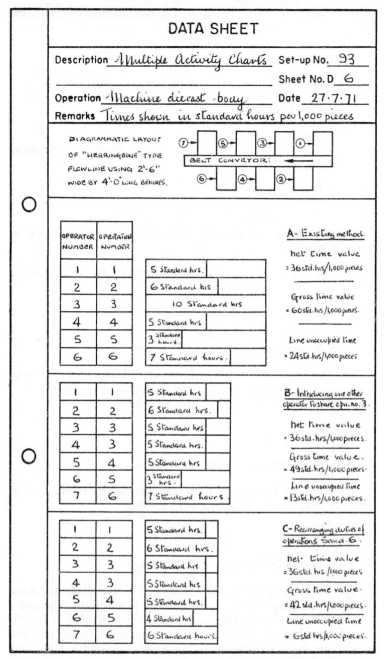

Fig 35 Schematic layout and multiple activity charts which show methods of reducing unoccupied time on machining operations performed on a small die cast component

The gross time value is now—

$$G = nl$$
$$= 7 \times 7$$
$$= 49 \text{ hours per } 1,000 \text{ pieces}$$

Since the net time value remains unchanged,

$$UT = G - N$$
$$= 49 - 36$$
$$= 13 \text{ standard hours per } 1,000 \text{ pieces}$$

(*g*) Unoccupied time was further reduced by rearranging the pattern of work, allocating some of it from operation No 6 (linish ends), to Operation 5 (linish face). The result of this reallocation was—

Operation	Time values (standard hours per 1,000 pieces)
5 Linish face and ends	4·0
6 Linish sides only	6·0

Under these new conditions there are two identical lead operations (Nos 2 and 6), both of these having the value of 6 standard hours per 1,000 pieces (Fig 35, diagram C).
The gross time value is now—

$$G = nl$$
$$= 7 \times 6$$
$$= 42 \text{ standard hours per } 1,000 \text{ pieces}$$

and the unoccupied time is—

$$UT = G - N$$
$$= 42 - 36$$
$$= 6 \text{ standard hours per } 1,000 \text{ pieces}$$

(*h*) By these two modifications, the unoccupied time has been reduced from 24 to 6 standard hours per 1,000 and the gross time value from 60 to 42 standard hours per 1,000, which is a reduction in cost of—

$$\frac{60 - 42}{60} \times 100 = 30\%$$

(*j*) The original operation efficiency was—

$$E = \frac{N}{G} \times 100$$

$$= \frac{36}{60} \times 100$$

$$= 60\%$$

and the new operation efficiency was—

$$E = \frac{36}{42} \times 100$$

$$= 86\%$$

A site plan of the line in sketch form is also shown in Fig 35.

Interference losses

Sometimes, operators working as a team may be forced to wait because of random delays when two or more operators finish their particular work at the same instant. This will cause loss in output, and where it occurs, it should be evaluated and added as an allowance to the standard time.

The phenomenon is more common on multiple machine work and is discussed at length in the next chapter.

A procedure for measuring the extent of this loss where it occurs on team work is given below—

(*a*) Determine the unoccupied time for the team per unit of production (i.e. per piece, per 100 pieces, per lb, etc.) by adding together all the unoccupied times per individual operation, with respect to the governing operation.

(*b*) Carry out a production study on the team. Use rated activity sampling or analytical observation (see Chapter 8). Record the output during the study period.

(*c*) Determine the performance of the team as a whole by taking the average rating and adjusting this by the relaxation allowance actually taken (see Chapter 7). Allow the team the unoccupied time when making this calculation.

(*d*) Compare the difference between the actual and theoretical outputs. If the actual output is lower than the theoretical output, the difference is interference loss. Add this loss as an allowance to all the work values, including the governing time.

10

The measurement of process-controlled work

Definitions

Process-controlled operations are those which are performed mainly by an automatic or semi-automatic machine or other device. Some work has usually to be done before the automatic cycle can commence or continue to run. This is known as *outside work*. Other, necessary work which can be performed whilst the process is continuing is called *inside work*, the total cycle time being the sum of the *process-controlled time* and the outside work.

These principles can be illustrated by considering machining of a small casting on a lathe fitted with an automatic traverse. The processes of loading the work to the chuck, moving the cutting tools to and from the surfaces to be machined and unloading the work after completion, can only be carried out when the automatic traverse is not operating; whereas it is possible to do such work as removing burrs and imperfections, procurement and disposal of castings, etc., whilst the machining is taking place. The complete operation can be represented by a *multiple activity chart* as shown in Fig 36, diagram A showing a typical cycle as it occurs, and diagram B the same operation with all the outside and inside work indicated as two single blocks. Here it can be seen that *unoccupied time* will occur per cycle by an amount which is equal to the difference between the process-controlled time and the inside work.

Another example is shown on the *multiple activity chart* for making tea in Fig 33. The original method of doing this operation meant a process-controlled time of 6 minutes, being made up of

137

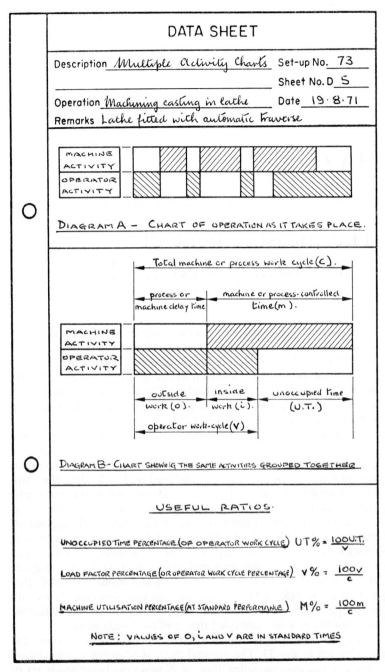

Fig 36 Multiple activity charts representing the machining of a small casting on a lathe showing the essential values and ratios

the 5 minutes time necessary to boil the kettle and the 1 minute to brew the tea. Work which can be performed within these cycles (i.e. the inside work) amounts to 3 minutes, and work which must be performed outside it (the outside work) totals 2 minutes. The unoccupied time per cycle is therefore 3 minutes, being the difference between the total process-controlled time and the inside work, and the total cycle time is 8 minutes, i.e. the process-controlled time plus the outside work.

Illustrative example

The measurement of process-controlled work can sometimes be a lengthy and complex procedure. An operation commonly performed in the textile industry during which yarn is wound from one type of bobbin to another prior to or during special processing of the material, incorporates a great deal of the problems encountered in this procedure. It is therefore quoted throughout the text in order to clarify the explanations.

The machines which are used consist mainly of two vertical spindles mounted above each other. One of these, approximately at shoulder height, can be set in motion or stopped by moving a small lever at the side. The other, which is at floor level, is stationary. Several of these spindles are arranged together in rows on the main machine structure which is known as a frame. Full and empty bobbins can be obtained from the storage trolleys moved conveniently near to the frame.

The operator is required to load full supply bobbins to the lower spindles, drawing the yarn through special guides on the machine and securing the end of it to empty "take-on" bobbins placed on the upper spindle, setting this in motion by moving the starting lever. If the tension of the yarn on the guides ceases due to breakage or completion of the winding process, the spindle will automatically stop. The operator will then either find the two ends, knot them together and re-start the spindle; or remove the fully wound take-on and empty supply bobbins and dispose of them to another storage trolley, replenishing the spindles with fresh work before doing so.

Exploration of existing working methods

The operations and equipment should be thoroughly investigated at an early stage until their significance is completely understood—

(*a*) Draw a site plan of the area to aid visualization of the processes and layout.

This need not be elaborate at this stage, a sketch roughly to scale often being sufficient.

139

(*b*) Carry out broad method study of the area. This should be standard practice in all work measurement applications. The intention should be to relieve local congestion and correct the more obvious faults by simple moving of benches and partitions, or by the provision of improved storage facilities, etc. No large-scale plant movements should be contemplated at this stage unless absolutely unavoidable.

In the *example*, it was discovered that the procurement and disposal of full and empty bobbins was cumbersome and lengthy due to poor arrangement of the conveyance trolleys and the absence of a satisfactory marshalling area. Slight improvements to the design of the trolleys and re-layout of the area to provide adequate parking facilities were therefore put in hand immediately.

(*c*) Check the machine operating characteristics. This is a most essential preliminary. Makers' handbooks, existing specifications, technical staff or other authorities may have to be consulted until completely satisfied that the process is being carried out to the best advantage. This would include checking any limiting time cycles such as cure times on rubber and plastic moulding, immersion times on metal plating, cooling cycles on metal casting, feed and speed times on metal cutting, machinery speeds, etc. Direct observation should always be used in spite of any helpful evidence which may be supplied in good faith by other authorities.

Spindle speeds should be checked with a revolution counter and a stop watch, the revolutions per minute being calculated afterwards. Operating speeds of machine slides should similarly be checked over a measured length. Where very high speeds exist, it may be necessary to use a stroboscopic disc or strobo-flash, particularly where the extra load imposed by a revolution counter might slow the machine. Tachometers are not generally recommended because of the difficulty of obtaining true average speeds by their use. Lathes or similar machines fitted with gear boxes should be checked throughout the whole range of their operation. Details of these observations should then be recorded on *work study observation sheets* and the result summarized on a *work study top sheet*, the whole being allocated a study number and filed for reference. Correct settings of machine speeds, process-operating cycles, etc., should then be established in consultation with competent technical staff supported by actual experiment if necessary. Any essential changes should then be reported to management to ensure that they are put into hand promptly.

Changes may not only apply to the machinery and process itself but also to the material. Great caution should be exercised throughout to ensure that the technical aspect of the process is not disturbed. This is why the technical authorities should be actively consulted at all stages.

In the *example*, spindle speeds were checked on each frame by revolution counter and stop watch. This revealed a wide variance in spite of the fact that all the machines were of the same make and model. Experiments carried out with the assistance of the technical manager and the machine makers' handbook established a standard speed of 1,180 revolutions per minute, this giving satisfactory quality standards consistent with safe operating conditions. Management agreed to take the necessary action, and new pulleys were made and fitted to some of the machines.

It is significant to mention that this action alone tended to increase the potential machine utilization by more than 5 per cent.

Process time study

In some operations, a process controlled time may vary according to the product. Such conditions occur on lathes, milling machines, etc., where varying diameters, lengths, depths of cut, type of material, class of finish, etc., may mean variation of cutting times. Other examples are in textile machines, where the size of yarn, weight of material wound, etc., will influence the machine-controlled time, or on die-casting, plastic injection moulding, etc., where the shape and size of the product will alter the injection and cooling cycles. In many instances, these details can be recorded simultaneously with time studies taken on the operator, but in other cases special studies may have to be taken separately.

The purpose of the investigation at this point is to attempt to discover relationships between these variations and the product itself, which can then be used to calculate them without further need for direct observation. It can be set out as follows—

(*a*) Decide the extent of the conditions and arrange a study programme around this.

This consists of broadly analysing the range of products and processes involved and listing those particular items which are regarded as most significant to the investigation. For example, if capstan lathes were under review, it may be decided that study of the largest and smallest parts together with one or two inter-mediate sizes on a range of different materials may be sufficient.

(*b*) Using a stop watch, a study board and a supply of *work study observation sheets* (Fig 9), take the process-controlled time over several cycles of the machine or process. Ensure that all conditions are set correctly and the operation is being performed properly before doing so and record all relevant details of the

141

equipment, speeds, material specifications, etc., as well as any delays or stoppages of cycle. For example, if timing the machining of a given length and diameter of metal, the specification of the metal, speed and feed times, condition of the cutting tools, standard of finish, depth of cut, type of machine, etc., should all be carefully observed, and the times when the cutting process ceases for any reason recorded on the *observation forms*.

If the cycle times are of very small duration, several of them may have to be observed together, individual times being obtained by use of the accumulative or differential timing methods described in Chapter 2.

(*c*) Summarize the results of each study on a *work study top sheet* (Fig 13) in a similar manner to that described in Chapter 3.

(*d*) Record details of each study on *data sheets* (Fig 25). Determine the relationships between the product, material, machine, and the process-controlled time. This follows a similar procedure to that discussed in Chapter 6. Each characteristic of the material, process, machine, and product must be very carefully considered to discover which of these cause corresponding variations in the time cycles. They are then set out in either tabular or graphical form to test these theories until a firm relationship is established. Machining times will tend to vary directly with the amount of metal removed, the class of finish, type of material and the methods of machining.

Cooling cycles on casting and plastic moulding processes tend to vary according to weight and surface area. Surface area can alternatively be expressed as a function of complexity of shape, length and weight. For example, a piece of metal of say 10 lb weight will tend to cool more rapidly if its maximum length in any direction is say, 5 feet as opposed to 1 foot, because it will have a greater surface area. Also, two castings of the same weight will cool at different rates if one is of a more intricate shape than the other as the more complex one would present more surface area and therefore cool more quickly. Cooling is also directly proportionate to weight assuming similar shapes. It can also be influenced by the addition of cooling water channels in the tools or machine used as well as extra amounts of metal caused by the addition of risers and vents to avoid the formation of air pockets and porosity.

In the *example*, the process-controlled times (i.e. the winding times per lb of yarn), were found to vary in different conditions. The procedure used to determine relationships of these to the specification of the yarn was as follows—

(*a*) 5 empty "take-on" bobbins were weighed on a chemical balance, each being numbered 1 to 5 and marked with their respective weights.

(*b*) An operator was requested to load the bobbins to marked spindles adjacent to each other, fixing the thread to each in turn at approximately half-minute intervals and starting the spindles in motion.

(*c*) The stop watch was started at the precise moment when the first spindle was set in motion and left running continuously until the yarn on the fifth bobbin was completely wound, the reading at which any of the spindles were started or stopped being recorded on a *work study observation sheet* of the type shown in Fig 9.

(*d*) The full take-on bobbins were again weighed on the chemical balance and both full and empty weights recorded on the study.

(*e*) Net winding times for the five bobbins were obtained by subtracting the total non-winding intervals from the elapsed time of the study.

(*f*) The net weight of yarn wound was also obtained by subtracting the weight of the empty bobbins from the full ones.

(*g*) After taking such studies on various yarns, results were tabulated as follows—

t = net winding time per take-on bobbin.
w = net weight of yarn wound.
d = denier of yarn.
f = number of folds of yarn.
s = spindle speed (rpm).

Since the denier of the yarn is proportionate to its actual diameter and the number of folds or twists also increases this diameter in proportion, then

diameter of yarn $\propto$ df.

If both the speed and the amount of yarn wound were constant, the winding time would increase as the diameter of yarn decreases.

or $\qquad t \propto 1/df$ (where w and s are constant).

Also, as the speed of the spindle increases, the winding time decreases

or $\qquad t \propto 1/s$ (where df and w are constant).

The winding time would also increase proportionate to the amount of yarn wound,

or $\qquad t \propto w$

143

Therefore, if w, df and s are all subject to variation then

$$t \propto w/dfs$$

Values of t were plotted against values of w/dfs. The result was a straight line graph which did not pass through the origin.

(*h*) By taking two points from the graph developed under (*g*) and equating these values as simultaneous equations using the straight line formula

$$t = \frac{Kw + k}{dfs}$$

values of the constant K and k were evaluated for the constant spindle speed of 1,180 rpm to give a formula of

$$t = \frac{13130w + 1152}{df}$$

and since

$$T = \frac{t}{w}$$

then

$$T = \frac{13130w + 1152}{dfw}$$

where $\quad T =$ the rate of winding in minutes per lb per spindle.

Thus it was possible to obtain rates of winding for any type of yarn and net weights wound from this formula without the necessity for further observations.

Numerical example of process time study

A numerical example is now given from the formula developed in the previous section. This will be used for further calculations later on in the chapter.

$$\text{yarn specification} = \text{2-fold 205 denier}$$

$$\text{net weight of yarn to be wound} = 1 \cdot 46 \text{ lb}$$

$$\text{spindle speed} = 1180 \text{ rpm}$$

$$\text{then the rate of winding } T = \frac{13130w + 1152}{dfw}$$

$$= \frac{(13130 \times 1 \cdot 46) + 1152}{205 \times 2 \times 1 \cdot 46}$$

$$= 34 \text{ mins per lb per spindle}$$

Determination and checking of net standard times

Standard times for process-controlled work should not be considered as complete until they have been adjusted to allow for unoccupied time or similar losses. Net standard times are therefore determined and checked as described in Chapters 3 to 7, the activities of the machine or process being ignored at this stage.

In the *example*, the following procedure was used—

(*a*) Elemental standard times were determined.

The operation was time studied when different sizes of yarn were being wound and the data developed into elemental standard times segregated into inside and outside work categories, using the *collection of elements sheet* (Fig 22) for the purpose—

(i) Standard times per take-on bobbin.
These were of constant duration for all sizes of yarn, and there was only one type of take-on bobbin used. The values were 0·605 and 0·130 standard minutes for the inside and outside work respectively.

(ii) Standard times per supply bobbin.
Four different types of supply bobbin were in use and there were different time standards for each, the choice being regulated by the previous processes and size of the yarn. The values I and O (for the inside and outside work respectively) were tabulated against the respective yarn specifications.

(iii) Standard times per repair of thread broken B.
Occasionally the thread would break, requiring the operator to find the two ends and repair it. All this work had to be done outside the process cycle. The times varied, a smooth curve being obtained when they were plotted against the denier of the yarn. A *graph sheet* (Fig. 19) was used for this purpose and the values extracted from it, tabulated for all sizes of yarn likely to be processed.

(*b*) A formula was developed from which provisional standard times for the inside, outside and total work per lb could be calculated for the different sizes and grades of yarn wound—

(i) The standard times per take-on and supply bobbins could be converted to standard times per lb by dividing each by the weight of the finished yarn wound w. This weight was dependent on the denier and type of yarn. The resultant times (i.e. $0·605/w$; $0·130/w$; I/w and O/w) were all evaluated and tabulated against the various yarn specifications used.

(ii) The frequency of repairing a broken thread was examined graphically and found to vary with the denier of the yarn, the nature of the previous processes, and the manufacturer of the raw material. Values of these frequencies F taken from the graphs were then tabulated against the appropriate denier and other yarn specifications.

(iii) The formula for calculating net standard times was now expressed as—

$$i = \frac{0 \cdot 605 + I}{w} + FB$$

$$o = \frac{0 \cdot 130 + O}{w}$$

$$v = i + o$$

$$= \frac{0 \cdot 735 + I + O}{w} + FB$$

where i = the inside work per lb of yarn
o = the outside work per lb of yarn
v = the total net standard time per lb

(c) Net standard times per lb of yarn were computed from the formula. An example of one of these is given below. It will be used for further calculation later on in the chapter.

yarn specification	= 2 fold 205 denier
net weight of yarn to be wound	= 1·46 lb
inside work per supply bobbin I	= 0·120 standard minutes
outside work per supply bobbin O	= 0·074 standard minutes
standard time per repair of thread broken B	= 0·368 standard minutes
number of breaks per lb F	= 0·110

then the inside work—

$$i = \frac{0 \cdot 605 + I}{w} + FB$$

$$= \frac{0 \cdot 605 + 0 \cdot 120}{1 \cdot 46} + (0 \cdot 110 \times 0 \cdot 368)$$

$$= 0 \cdot 497 + 0 \cdot 041$$

$$= 0 \cdot 538 \text{ standard minutes per lb}$$

and the outside work—

$$o = \frac{0 \cdot 130 + O}{w}$$

$$= \frac{0 \cdot 130 + 0 \cdot 074}{1 \cdot 46}$$

$$= 0 \cdot 140 \text{ standard minutes per lb}$$

and the total net standard time—

$$v = i + o$$
$$= 0 \cdot 538 + 0 \cdot 140$$
$$= 0 \cdot 678 \text{ standard minutes per lb}$$

(*d*) Net standard times per lb of yarn computed from the formula were checked from existing time studies taken over at least two hours. A numerical example of checking two of these values using one study is given below—

(i) The following details were extracted from the studies—

y = weight of yarn produced of each type (120 lb of 2 fold 205 denier and 182 lb of 2 fold 100 denier)
e = the effective time of the study (201 minutes)
r = the average rating (96)
R = the average relaxation allowance included in the standard times (14%)
v = the net standard times per lb of yarn. (2 fold 205 denier = 0·678 standard minutes/lb and 2 fold 100 denier = 0·789 standard minutes/lb)

(ii) The total value of measured work determined from—

$$W = v_1 y_1 + v_2 y_2, \text{ etc.}$$
$$= (0 \cdot 678 \times 120) + (0 \cdot 789 \times 182)$$
$$= 81 \cdot 4 + 143 \cdot 6$$
$$= 225 \text{ standard minutes}$$

(iii) The theoretical time taken to produce the above amount of measured work, assuming that 14 per cent *RA* was taken and that there were no unoccupied time or machine interference losses from—

$$H = \frac{(100 + R)e}{100}$$

$$= \frac{(100 + 14)\,201}{100}$$

$$= 229 \text{ minutes}$$

(iv) The operator performance calculated from—

$$= \frac{100\,W}{H}$$

$$= \frac{100 \times 225}{229}$$

$$= 98 \text{ BSI}$$

(v) The operator performance was compared with the average rating. If they were within $\pm 2\frac{1}{2}$ per cent of each other, the net standard times were regarded as being valid—

In the *example*—

$p\ =\ 98$ BSI

and

$r\ =\ 96$ BSI

Since these differed by less than $2\frac{1}{2}$ per cent, the net standard times were regarded as being valid.

Determination of unoccupied time on a single machine or process

Unoccupied time can be expressed in various ways, e.g. per cycle, per hour, per unit of output (i.e. per piece, per unit weight, length, volume, etc.), or as a percentage.

Two of these are considered, UT per cycle and percentage UT, the latter being the most important since it can be simply converted to any of the others.

Consider the diagram Fig 36.

Let the process-controlled time per cycle $= m$
the inside work per cycle $= i$
the outside work per cycle $= o$

then the unoccupied time per cycle—

$$UT\ =\ m - i$$

and the total cycle time—

$$C\ =\ m + i$$

the operator work cycle per total cycle—

$$v\ =\ i + o$$

the percentage unoccupied time of the operator work cycle—

$$UT\% = \frac{100UT}{v}$$

$$= \frac{100UT}{i + o}$$

$$= \frac{100(m - i)}{i + o}$$

the operator work cycle percentage of the total cycle time (or load factor)—

$$v\% = \frac{100v}{C}$$

$$= \frac{100v}{m + o}$$

the net standard time per unit of output—

$$V = \frac{v}{q}$$

where q = the number of pieces, lb, etc., produced per cycle. the standard time per unit of output inclusive of unoccupied time—

$$ST = \frac{100 + UT\%V}{100}$$

the machine utilization index—

$$M\% = \frac{100m}{C}$$

$$= \frac{100m}{m + o}$$

Reduction of unoccupied time

Unoccupied time on machine or process-controlled operations can be reduced by the following means—

(*a*) Classification of as much work as possible to inside work. This has the combined effect of reducing the overall cycle time as well as reducing the unoccupied time (see Fig 37).

(*b*) Arrangement of more duties to be performed during the unoccupied time period, such as oiling or greasing of machines, etc., procurement and disposal of materials, etc. (see Fig 37).

149

6

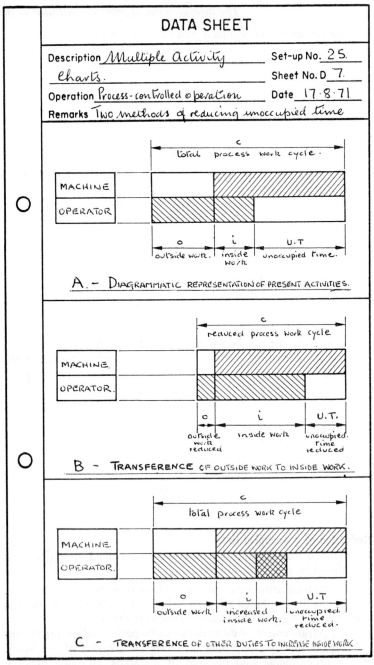

DATA SHEET

Description _Multiple Activity_   Set-up No. 25.
Charts.   Sheet No. D 7.
Operation _Process-controlled operation_   Date 17·8·71
Remarks _Two methods of reducing unoccupied time_

A. — DIAGRAMMATIC REPRESENTATION OF PRESENT ACTIVITIES.

B — TRANSFERENCE OF OUTSIDE WORK TO INSIDE WORK.

C — TRANSFERENCE OF OTHER DUTIES TO INCREASE INSIDE WORK

Fig 37 Multiple activity charts on a process-controlled operation
showing methods of reducing unoccupied time

Such extra work is in fact inside work, the unoccupied time being calculated as before but using this increased value, i.e.—

$$UT \text{ per cycle} = c - i$$

$$\text{and } UT\% = \frac{100(c - i)}{i + o}$$

using the new values of i.

(*c*) Allocation of more than one machine to an operator. This is known as *multiple machine work* (see Fig 38). A different method of calculating the unoccupied time is necessary, and extra losses in the form of machine interference may result. These are now discussed in more detail.

Determination of unoccupied time on multiple machine work

The possibility of arranging for an operator to manage a second machine and thereby reduce unoccupied time to a minimum without serious loss of machine utilization will depend on whether the sum of the operator cycle times does not exceed the sum of the unoccupied times. For example, if two machines were under consideration as follows (standard minutes are henceforth abbreviated to *SM*'s)—

	1st machine	*2nd machine*	*Total*
UT per cycle	10 minutes	15 minutes	25 minutes
operator cycle time	15 *SM*'s	10 *SM*'s	25 *SM*'s
Total cycle time at standard performance	25 minutes	25 minutes	50 minutes

the sum of the unoccupied times and operator cycle times are the same. It is therefore theoretically possible to eliminate unoccupied time if an operator works both machines at standard performance. Consider also the following example—

	1st machine	*2nd machine*	*Total*
UT per cycle	12 minutes	19 minutes	31 minutes
operator cycle time	16 *SM*'s	11 *SM*'s	27 *SM*'s
Total cycle time at standard performance	28 minutes	30 minutes	58 minutes

151

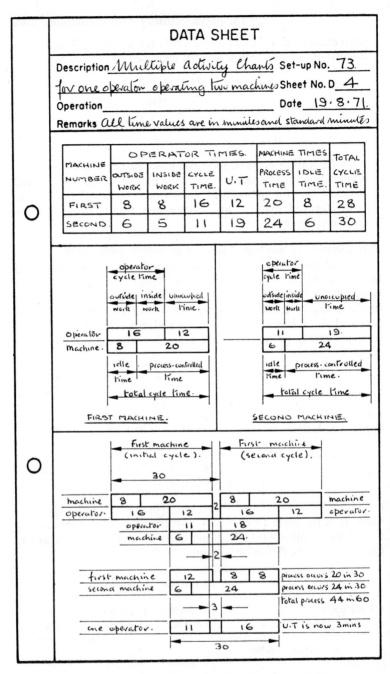

DATA SHEET

Description *Multiple Activity Charts* Set-up No. 73
for one operator operating two machines Sheet No. D 4
Operation _____ Date 19.8.71.
Remarks *All time values are in minutes and standard minutes*

MACHINE NUMBER	OPERATOR TIMES.				MACHINE TIMES		TOTAL CYCLE TIME
	OUTSIDE WORK	INSIDE WORK	CYCLE TIME.	U.T	PROCESS TIME	IDLE TIME.	
FIRST	8	8	16	12	20	8	28
SECOND	6	5	11	19	24	6	30

Fig 38 Multiple activity charts showing the effect of allocating two machines to one operator

In this case, the unoccupied time on any one machine exceeds the operator cycle time on the other; it is therefore possible to allocate both machines to one operator, although there would still be some unoccupied time remaining.

	1st machine	2nd machine	Total
UT per cycle	12 minutes	19 minutes	31 minutes
Operator cycle time on the other machine	11 minutes	16 minutes	27 minutes
Differences at standard performance (net UT)	+1 minute	+3 minutes	+4 minutes

As can be seen from Fig 38, the unoccupied time would be equal to the larger net *UT* value (i.e. 3 minutes) and this would occur every time the longest cycle of 30 minutes is completed, i.e.—

$$UT\% = \frac{100 \times \text{largest value of net } UT}{\text{longest total cycle time} - \text{largest value of net } UT}$$

and for the example in Fig 38—

$$UT\% = \frac{100 \times 3}{30 - 3}$$
$$= 11 \cdot 1\%$$

also,

$$M\% = \frac{100 \times \text{the sum of the process-controlled times}}{n \times \text{longest total cycle time}}$$

and for the example in Fig 38—

$$M\% = \frac{100(24 + 20)}{2 \times 30}$$
$$= 73 \cdot 4\%$$

where

$M\% =$ the machine utilization percentage
$n =$ the number of machines allocated (in this example $n = 2$).

and

$m =$ the process-controlled times (in this example 24 and 20 minutes respectively).

153

Where more than two machines are allocated consideration must be given to the activity during the unoccupied time period on each individual machine. In the example shown in Fig 39, the three machines might be operating as follows—

	1st machine	2nd machine	3rd machine	Total
UT per cycle	12 minutes	16 minutes	20 minutes	48 minutes
operator cycle time	8 *SM*'s	9 *SM*'s	4 *SM*'s	21 *SM*'s
Total cycle time at standard performance	20 minutes	25 minutes	24 minutes	69 minutes

If the operator cycle time on any two machines was performed during the unoccupied time of a third, the following conditions would result—

	1st machine	2nd machine	3rd machine	Total
UT per cycle	12 minutes	16 minutes	20 minutes	48 minutes
Sum of the operator cycle times on the other two machines	(9 + 4) =13 *SM*'s	(8 + 4) =12 *SM*'s	(8 + 9) =17 *SM*'s	42 *SM*'s
Difference at standard performance (net *UT*)	−1 minute	+4 minutes	+3 minutes	+6 minutes

Referring to Fig 39, it will be seen that the unoccupied time per cycle is equal to the largest value of the net *UT* in the table above, and this would again occur every time the longest cycle of 25 mins. is completed, i.e.—

$$UT\% = \frac{100 \times \text{largest value of net } UT}{\text{longest total cycle time—largest value of net } UT}$$

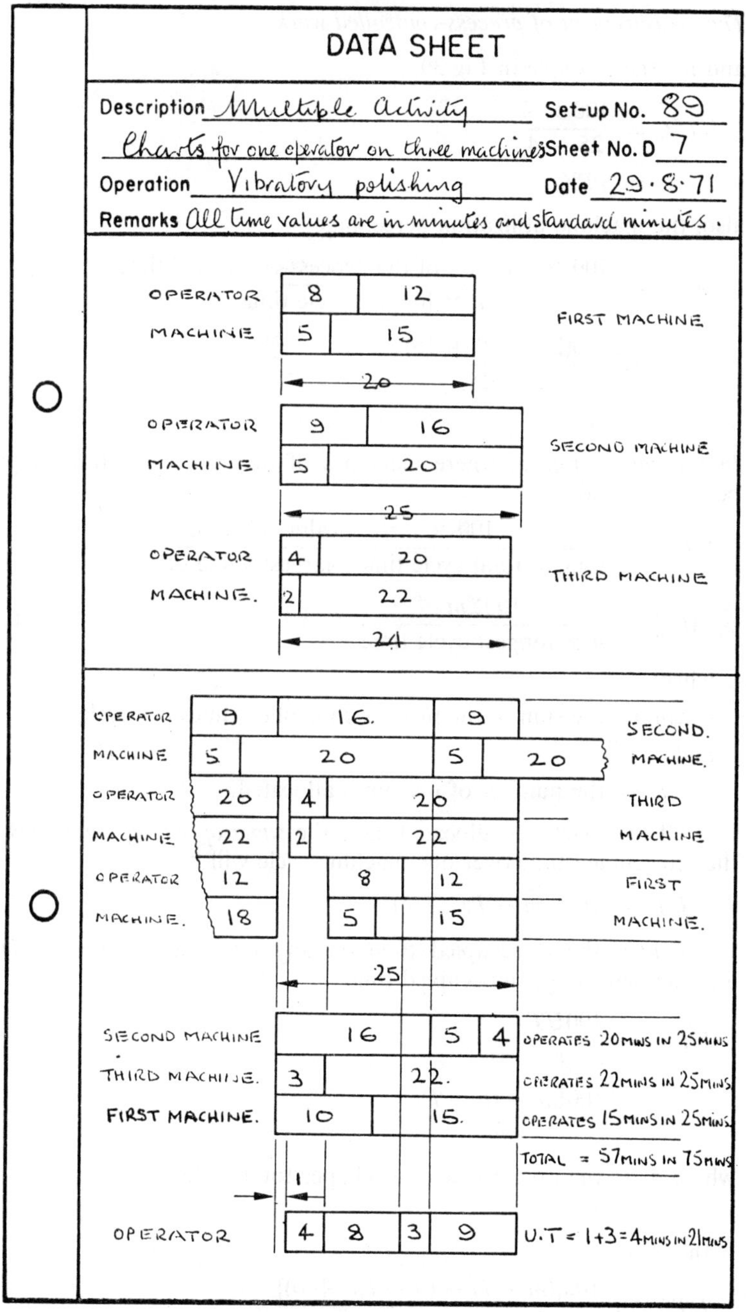

Fig 39 Multiple activity charts showing the effect of allocating
three machines to one operator

and for the example in Fig 39—

$$UT\% = \frac{100 \times 4}{25 - 4}$$

$$= 19\%$$

also, the machine utilization percentage—

$$M\% = \frac{100 \times \text{the sum of the process-controlled times}}{n \times \text{longest cycle time}}$$

$$= \frac{100(22 + 20 + 15)}{3 \times 25}$$

$$= 76\%$$

In general, therefore, where machines of different cycle times are being allocated—

$$UT\% = \frac{100 \times \text{largest value of net } UT}{\text{longest total cycle time—largest value of net } UT}$$

$$M\% = \frac{100\Sigma m}{n \times \text{longest cycle time}}$$

where

Σm = the sum of the process-controlled times per cycle

and

n = the number of machines allocated.

If all the machines allocated have the *same operating conditions*, the unoccupied time over one machine cycle will be—

$$UT = ut - (n - I)(i + o)$$

where ut = the unoccupied time on any one machine ($= c - i$) and the percentage unoccupied time—

$$UT\% = \frac{100UT}{nv}$$

$$= \frac{100\{ut - (n - I)v\}}{nv}$$

where v = the operator work cycle per total cycle—

$$v = i + o$$

therefore—

$$UT\% = \frac{100\{(m - i) - (n - I)(i + o)\}}{n(i + o)}$$

156

where

m = the process-controlled time per cycle
i = the inside work per cycle
o = the outside work per cycle
n = the number of machines allocated.

Also where all the machines have the *same operating conditions*, the maximum number which can be allocated theoretically to reduce unoccupied time to a minimum without serious effect on machine utilization is therefore—

$$n = \frac{ut}{v} + 1$$

$$= \frac{m - i}{i + o} + 1$$

Calculation of machine interference allowances

Allocation of more than one machine to an operator should tend to reduce or eliminate unoccupied time. There is always the possibility, however, that one or more of the machines will complete their cycles before others, needing attention, can be re-started. Such a condition, known as *machine interference*, can be an independent source of loss.

Sometimes, its effect is very small, in which case the value is more conveniently determined by using direct observation and study; but in other instances, particularly where large numbers of machines can be allocated, approximate calculations should first be carried out before exact evaluation is made from studies taken later on.

The procedure explained in this section deals with these calculations, and is useful for the determination of theoretically optimum conditions which can then be tested and studied in practice.

(*a*) Determine the minimum allocation of machines at standard performance which would theoretically eliminate unoccupied time.

This is carried out using the formula developed previously for calculating the number of machines, i.e.—

$$n = \frac{m - i}{i + o} + 1$$

Let a = the maximum value of n (i.e. that condition which would theoretically eliminate unoccupied time).

then $a = \dfrac{m - i}{i + o} + 1$

To explain this, the value of *a* in the *example* would be—

$$a = \frac{34 - 0\cdot538}{0\cdot538 + 0\cdot140} + 1$$

$$= 50\cdot4 \text{ machines}$$

Where *m* (the process-controlled time) is a winding time of 34 minutes.

> *i* (the inside work) is 0·538 *SM*'s
> *o* (the outside work) is 0·140 *SM*'s

(*b*) Determine the minimum allocation of machines at standard performance which would theoretically eliminate unoccupied time if there was no outside work, from—

$$M = m/i$$

In the example, this would be—

$$M = \frac{34}{0\cdot538}$$

$$= 63\cdot2 \text{ machines}$$

(*c*) Determine the following for various values of *n* commencing at the maximum value *a* and progressing downwards. To illustrate the method of doing this, several values have been calculated for allocation of machines *n* from 50·4 to 46, and these are shown in the table Fig 40—

(i) The average number of machines which will probably be stopped at any instant due to machine interference from the following formula, which is based on the queuing theory—

$$s = \frac{n}{M - n}$$

In the example—
where *n* = *a*, the value of *s* would be—

$$s = \frac{50\cdot4}{63\cdot2 - 50\cdot4}$$

$$= 3\cdot9 \text{ machines}$$

also where *n* = 50 machines—

$$s = \frac{50}{63\cdot2 - 50}$$

$$= 3\cdot8 \text{ machines, etc.}$$

Symbol	Formula	Details	Values					
n	max. value of $n = a$ $a = 50.4$	Number of machines allocated	50·4	50	**49**	48	47	46
s	$\dfrac{n}{M-n}$ where $M = 63.2$	Average number of machines stopped due to interference	3·9	3·8	**3·5**	3·2	2·9	2·7
ut	$a - n$	Average number of machines stopped due to under-allocation of machines	—	0·4	**1·4**	2·4	3·4	4·4
S	$s + ut$	Average number of machines stopped due to under-allocation and interference	3·9	4·2	**4·9**	5·6	6·3	7·1
N	$n - s$	Average number of machines running	46·5	46·2	**45·5**	44·8	44·1	43·3
$S\%$	$\dfrac{100S}{N}$	Per cent UT + machine interference	8·3	9·1	**10·8**	12·5	14·3	16·4
$M\%$	$\dfrac{100N}{n}$	Machine utilization index	92·4	92·4	**93·0**	93·5	93·8	94·2
$L\%$	$\dfrac{100N}{a}$	Labour utilization index	92·4	92·5	**91·2**	89·7	88·3	86·7
MC	$\dfrac{100k}{M\%}$ where $k =$ cost per machine hour $= 2p$	Machine cost in pence per hour	2·17	2·17	**2·15**	2·14	2·13	2·12
LC	$\dfrac{100k}{nL\%}$ $K =$ std direct labour cost/hr $= 40p$	Labour cost in pence per hour	0·86	0·87	**0·88**	0·93	0·94	1·00
TC	$MC + LC$	Total cost in pence per hour	3·03	3·04	**3·03**	3·07	3·07	3·12

Fig 40 Table showing steps in the calculation of machine interference allowances

(ii) The number of machines under-allocated, and which will cause unoccupied time to occur because of this from—

$$ut = a - n$$

In the example—
where $n = a = 50\cdot4$ machines, the value of ut would be—

$$ut = 50\cdot4 - 50\cdot4$$
$$= 0 \text{ machines}$$

also where $n = 50$ machines—

$$ut = 50\cdot4 - 50$$
$$= 0\cdot4 \text{ machines, etc.}$$

(iii) The total average number of machines not available for work and lost to the operator due to the combined effect of unoccupied time and machine interference from—

$$S = s + ut$$

In the example—
where $n = a = 50\cdot4$ machines, the value of S would be—

$$S = 3\cdot9 + 0$$
$$= 3\cdot9 \text{ machines}$$

also where $n = 50$ machines—

$$S = 3\cdot8 + 0\cdot4$$
$$= 4\cdot2 \text{ machines, etc.}$$

(iv) The average number of machines expected to be running at any instant at standard performance from—

$$N = n - s$$

In the example—

where $n = a = 50\cdot4$ machines—

$$N = 50\cdot4 - 3\cdot9$$
$$= 46\cdot5 \text{ machines}$$

also where $n = 50$ machines—

$$N = 50 - 3\cdot8$$
$$= 46\cdot2 \text{ machines, etc}$$

(v) The total percentage which would have to be allowed to an operator to compensate for losses due to the combined effect of unoccupied time and machine interference from—

$$S\% = 100S/N$$

In the example—

where $n = a = 50{\cdot}4$ machines, the values of $S\%$ would be—

$$S\% = \frac{100 \times 3{\cdot}9}{46{\cdot}5}$$

$$= 8{\cdot}4\%$$

also where $n = 50$ machines—

$$S\% = \frac{100 \times 4{\cdot}2}{46{\cdot}2}$$

$$= 9{\cdot}1\%, \text{ etc.}$$

(vi) The machine utilization index from—

$$M\% = 100N/n$$

In the example—

where $n = a = 50{\cdot}4$ machines—

$$M\% = \frac{100 \times 46{\cdot}5}{50{\cdot}4}$$

$$= 92{\cdot}4\%$$

also where $n = 50$ machines

$$M\% = \frac{100 \times 46{\cdot}2}{50}$$

$$= 92{\cdot}4\%, \text{ etc.}$$

(vii) The labour utilization index from—

$$L\% = 100N/a$$

In the example—

where $n = a = 50{\cdot}4$ machines—

$$L\% = \frac{100 \times 46{\cdot}5}{50{\cdot}4}$$

$$= 92{\cdot}4\%$$

also where $n = 50$ machines—

$$L\% = \frac{100 \times 46 \cdot 2}{50 \cdot 4}$$

$$= 91 \cdot 8\% \text{ etc.}$$

The values of L per cent are the effective performances of the operator when he is working at standard performance wherever it is possible for him to do so, bearing in mind that he is prevented from working during the unoccupied time and machine interference periods.

(viii) The cost of running the machines per hour from—

$$MC = \frac{100k}{M\%}$$

where $k =$ the cost per machine hour at 100 per cent utilization.

In the example—

where $n = a = 50 \cdot 4$ machines—

$$MC = 100 \times 2/92 \cdot 4$$

$$= 2 \cdot 17 \text{ pence per hour.}$$

also where $n = 50$ machines—

$$MC = 100 \times 2/92 \cdot 4$$

$$= 2 \cdot 17 \text{ pence per hour, etc.}$$

Note: In the example it is assumed that the total overhead cost per hour for running one machine at 100 per cent utilization was 2 pence.

(ix) The labour cost per machine hour from—

$$LC = \frac{100K}{nL\%}$$

where $K =$ the net standard direct labour cost per hour.

In the example—

where $n = a = 50 \cdot 4$ machines—

$$LC = \frac{100 \times 40}{50 \cdot 4 \times 92 \cdot 3}$$

$$= 0 \cdot 86 \text{ pence per machine hour}$$

also where $n = 50$ machines—

$$LC = \frac{100 \times 40}{50 \times 91 \cdot 8}$$

= 0·87 pence per machine hour, etc.

Note: In the example it is assumed that the net standard direct labour cost per hour is 40 pence.

(x) The total cost per machine hour from—

$$TC = MC + LC$$

In the example—

when $n = a = 50 \cdot 4$ machines—

$$TC = 2 \cdot 17 + 0 \cdot 86$$
= 3·03 pence per hour

also where $n = 50$ machines—

$$TC = 2 \cdot 17 + 0 \cdot 87$$
= 3·04 pence per hour, etc.

(*d*) Select the conditions which give the lowest total cost per hour and arrange to conduct a trial run with *n* number of machines.

In the *example* this will occur when $n = 49$ machines and the total cost is 3·03 pence per hour (since it is obviously impractical to run 50·4 machines, the other condition which gives the same total cost). See table Fig 40.

Determination of ancillary allowance

During certain processes, it is sometimes necessary to start an empty machine before it can be operated, and when the particular run is completed, allow it to return to its previous empty condition before changing over to a different product. A standard time must be calculated which is equivalent to the loss of production during this change-over period to compensate the operator. This is known as a machine ancillary time allowance.

In the *example*, these were calculated to allow for loss of production during the time the machines were standing when being changed over from one type of yarn to another, it being necessary to alter the tensioning devices on each machine spindle.

The loss of production was equal to—

$$\text{Ancillary allowance} = Ww$$

Where W = the provisional standard time per lb (0·678 *SM*'s)
and w = the weight of the finished yarn wound (1·46 lb).

163

The reason for this is that a complete cycle of the machine during which 1·46 lb of yarn could have been wound is lost. This would have taken 0·678 minutes for each lb wound. Therefore the loss in standard minutes per machine is—

$$\text{Ancillary allowance} = 0·678 \times 1·46$$
$$= 1·00 \text{ standard minute per machine (spindle)}$$

Analysis of operation under theoretically optimum conditions
The procedure at this stage is to test any proposed modifications to the process under actual conditions and finally to calculate the unoccupied time and machine interference losses. It can be carried out as follows—

    (*a*) Time study the operation under the new conditions.

    (*b*) Determine the effective operator performance and from this, the total measured work due to the operator.

    (*c*) Determine the actual or equivalent output of the machines or processes during the study period, and from this and the net standard time determine the net measured work produced.

    (*d*) Determine the combined unoccupied time and machine interference losses by comparison of the values obtained during steps (*b*) and (*c*).

    (*a*) *Time study the operation under the new conditions*
This consists of observing a special run of the machines under proposed new conditions. Time study can be carried out by either of three methods, the choice being dictated by the nature of the work. The success of such studies depends to a large extent on the co-operation of the operator and the lack of any disturbing influences such as waiting time or machine breakdown. Care should therefore be taken in the selection of a suitable operator and full support given by shop supervision to ensure that conditions are as truly representative as possible—

  (i) Take a time study of fairly long duration (say 4 to 8 hours). Time and rate all work elements. Take particular note of any idle period where work is not available (unoccupied time); and idle periods where work is available (idle time). Note the performance of the machines or processes using either activity sampling or direct study—with a second observer if necessary.

  (ii) Use activity sampling to check the machine performance and rated activity sampling to obtain the average rating, the unoccupied time and the idle time. This method has the added advantage that several different arrangements can be studied at a time by patrolling each bank of machines at random intervals. Against this must be set the large number of observations which may be necessary due to the ultimate accuracy needed.

(iii) Take a time study as in (i) above but record the output directly during a study period. It may be necessary to arrange for a special run of the machines in order to carry this out. The procedure for doing this is described later in this chapter under *alternative method of determining unoccupied time and machine interference losses.*

(*b*) *Determine the total measured work due to the operator*
This is carried out as follows—

(i) Extract information from the time or rated activity sampling studies. Values applying to the *example* are shown in brackets (see also Chapter 7—*checking procedures*)—

$$E = \text{the elapsed time} \qquad (284 \text{ minutes})$$
$$e = \text{the effective time} \qquad (220 \text{ minutes})$$
$$I = \text{the idle time} \qquad (23 \text{ minutes})$$
$$UT = \text{the unoccupied time} \qquad (21 \text{ minutes})$$
$$r = \text{the average rating} \qquad (101)$$
$$c = \text{the check time} \qquad (5 \text{ minutes})$$
$$l = \text{the waiting time} \qquad (10 \text{ minutes})$$
$$t = \text{the net study time}$$
$$= \{E - (c + l)\} \qquad (269 \text{ minutes})$$

(ii) Obtain the average relaxation allowance percentage included in the net standard times. In the *example* this was—
$$R = 14\%$$

(iii) Determine the overall relaxation factor percentage from—
$$F = 100 + R$$

In the *example* this was—
$$F = 100 + 14$$
$$= 114\%$$

(iv) Determine the percentage relaxation allowance taken during the study from—
$$R\% = \frac{100I}{e}$$

In the *example* this was—

$$R\% = \frac{100 \times 23}{220}$$

$$= 10 \cdot 5\%$$

(v) Determine the effective operator performance from—

$$p = \frac{Fr}{100 + R\%}$$

In the *example* this was—

$$p = \frac{114 \times 101}{110 \cdot 5}$$

$$= 104$$

(vi) Determine the total measured work due to the operator from—

$$V_1 = \frac{pt}{100}$$

In the *example* this was—

$$V_1 = \frac{104 \times 269}{100}$$

$$= 280 \; SM\text{'s}$$

(c) *Determine the net measured work produced*

(i) Calculate the percentage machine utilization index from the time or activity sampling study taken on the machine(s)—

$M\% =$ the machine running time as a percentage of the study time.

In the *example*, this was obtained from details taken from the study as follows—

Total number of observations taken during the study time   261
Numbers of these observations during which the machines were found to be running   250

$$M\% = \frac{100 \times 250}{261}$$

$$= 95 \cdot 8\%$$

where

$M\% =$ the percentage machine utilization index.

(ii) Calculate the output of the machines at 100 per cent machine utilization which would have occurred during the study period.

$$Y = nt \text{ (the rate of producing)}$$

Since the rate of producing is—

$$= \frac{\text{output per cycle of each machine}}{\text{the sum of the process-controlled times}}$$

$$Y = \frac{nt \text{ (the total output per cycle of } n \text{ machines)}}{\text{the sum of the process-controlled times}}$$

To illustrate this, take the example given in Fig 39, the details being—

Machine number	Process-controlled time	Output per cycle
1	20	$x_1 = 10$ units
2	22	$x_2 = 20$ units
3	15	$x_3 = 5$ units
	Total 57	

then, if there were no stoppages of the process, at a period equal to the sum of the process-controlled times (i.e. in the example every 57 minutes), the production would be as follows—

$$10 \text{ units of } x_1$$

$$20 \text{ units of } x_2$$

$$\text{and} \quad 5 \text{ units of } x_3$$

Since, however, there are three machines running together this output would be produced during a time equal to—

$$\frac{\text{the sum of the process-controlled times}}{3}$$

and for n machines—

$$\frac{\text{the sum of the process-controlled times}}{n}$$

In this *example* (Fig 39), for a given study time t of 300 minutes, the output at 100 per cent machine utilization index would be—

$$Y = \frac{nt \text{ (the total output per cycle of } n \text{ machines)}}{\text{the sum of the process-controlled times}}$$

$$= \frac{3 \times 300(10x_1; 20x_2; 5x_3)}{57}$$

$$= 15 \cdot 8 \ (10x_1; 20x_2; 5x_3)$$

$$= 158x_1; 316x_2 \text{ and } 79x_3$$

In the *example* of winding yarn, the output would be—

$$Y = \frac{nt}{T}$$

where T = the rate of winding in minutes per lb per spindle (or machines) calculated during process time study at a value of 34 minutes per lb per spindle.

n = the number of machines (49)

t = the study time (269 minutes)

therefore,

$$Y = \frac{49 \times 269}{34}$$

$$= 388 \text{ lb of yarn}$$

(iii) Calculate the actual output of the machine(s).
In the example Fig 39, given $M\%$ as 72·5 per cent then—

$$y = \frac{72 \cdot 5}{100} \ (158x_1; 316x_2; 79x_3)$$

$$= 114x_1; 229x_2; \text{ and } 57x_3.$$

In the *example* of winding yarn, this would be

$$y = \frac{M\%Y}{100}$$

$$= \frac{95 \cdot 8 \times 388}{100}$$

$$= 372 \text{ lb of yarn}$$

(in every case y = the actual output of the machine(s)).

(iv) Calculate the measured work actually produced by multiplying the output by the net standard time(s), i.e.—

$$V_2 = y_1x_1 + y_2x_2 + y_3x_3, \text{ etc.}$$

In the *example*, Fig 39, given standard times per unit of 0·9 SM's for x_1; 0·2 SM's for x_2 and 1·6 SM's for x_3 then—

$$V_2 = (114 \times 0·9) + (229 \times 0·2) + (57 \times 1·6)$$
$$= 103 + 46 + 91$$
$$= 240 \ SM\text{'s}$$

and in the *example* of winding yarn; this was—

$$V_2 = 0·678 \times 372$$
$$= 252 \ SM\text{'s}$$

(*d*) *Determine the combined unoccupied time and machine interference losses*

This is obtained by comparison of the total measured work due (V_1) with the total measured work produced (V_2) using the net standard times.

In the *example*, Fig 39, given an effective operator performance of 98 per cent, and a study time of 300 minutes then the total measured work due to the operator—

$$V_1 = \frac{pt}{100}$$
$$= \frac{98 \times 300}{100}$$
$$= 294 \ SM\text{'s}$$

and the combined unoccupied time and machine interference losses—

$$S\% = \frac{100(V_1 - V_2)}{V_2}$$
$$= \frac{100(294 - 240)}{240}$$
$$= 22·5\%$$

and in the example of winding yarn, this was—

$$S\% = \frac{100(280 - 252)}{252}$$
$$= 11·1\%$$

(This is within $2\frac{1}{2}$ per cent of the calculated value of 10·8 per cent.)

(*e*) *Alternative method of determining a combined unoccupied time and machine interference losses*

An alternative method of determining a combined effect of unoccupied time and machine interference can be carried out as follows,

169

where the process-controlled cycles are of relatively short duration and the length of the study sufficient to ensure that the output produced during it is truly representative of the effort exerted.

In the *example*, a study of approximately 8 hours' duration on an allocated load of 49 spindles gave the following results—

Weight of yarn wound = 636 lb
Measured work actually produced using the net standard times—

$$V_2 = 636 \times 0.678$$
$$= 431 \ SM\text{'s}$$

Other details were—

$t =$ the study time—467 minutes
$p =$ the effective operator performance calculated as previously described—110

Therefore, the total measured work due to the operator was

$$V_1 = \frac{110 \times 431}{100}$$
$$= 474 \ SM\text{'s}$$

and the combined unoccupied time and machine interference losses—

$$S\% = \frac{100(V_1 - V_2)}{V_2}$$
$$= \frac{100(474 - 431)}{431}$$
$$= 10.0\%$$

(*f*) *Adjustment of conditions if there is a large discrepancy between the theoretical and actual unoccupied time and machine interference losses*

In cases where there is large discrepancy between the calculated value of $S\%$ and the actual value, proceed as follows—

(i) Using the actual value of $S\%$, recalculate the value of M.
(ii) Carry out the complete procedure as given under "Calculation of Machine Interference Allowance" and determinate another optimum condition, and from this the final value of $S\%$. Check this new value under actual conditions if the discrepancy has been very large.

To illustrate this procedure in part, the re-calculation of M is carried out below for the *example*, assuming that this time the actual value of $S\%$ for 49 machines was 15 per cent.

$$S\% = \frac{100S}{N}$$

but $S = s + ut$
$$= s + 1\cdot4$$

and $N = n - s$
$$= 49 - s$$

since $S\% = 15\%$

substituting—

$$15 = \frac{100(s + 1\cdot4)}{49 - s}$$

and $s = \dfrac{735 - 140}{100 + 15}$

$$= 5\cdot17$$

now $s = \dfrac{n}{M - n}$

and $M = \dfrac{49(s + 1)}{s}$

$$= \frac{49(5\cdot17 + 1)}{5\cdot17}$$

$$= 58\cdot4$$

With this new value of M, other values of the total cost TC are calculated as before and optimum conditions again determined. If there has been a large discrepancy the new optimum may have to be checked by carrying out another practical test.

Determination of final standard times

Theoretically, the final standard times should remain net, as adding allowances in the form of unoccupied time, etc., destroys the concept of the work unit. Also, since unoccupied time is a source of loss, it should not tend to be forgotten by burying it in standard times. Machine interference allowances however cannot be reduced and therefore it is common practice to include them in the issued standard times.

Some managements do not wish to include UT at standard and it may be necessary to reduce the $UT\%$ as calculated beforehand. For example, if it is the policy to include such losses at 90 performance, then for a $UT\%$ of 10, the allowance to be added would be—

$$90\% \text{ of } 10\%$$
$$\text{or } 9\%$$

In the *example* the combined unoccupied time and machine interference allowance was fixed at 11 per cent, and it was decided to include the whole of this in the standard times, i.e.—

Standard time value for winding 2 fold 205 denier yarn

$$= \frac{0.678 \times (100 + S\%)}{100}$$

$$= \frac{0 \cdot 678 \times 111}{100}$$

$$= 0 \cdot 750 \text{ standard minutes per lb}$$

This was applicable to a full load of spindles which was set at 49.

11

Final presentation of data

Work specifications
Standard times must be qualified by a work specification before issue. Accompanying this should be a guarantee that no time value will be altered unless there is a change in the method of working—or there has been a genuine mistake. This is to provide a safeguard for both management and labour if the values are to be used for incentive payment.

Work specifications should be clearly and concisely written so that they can be easily understood by management and operators alike. Such features as make of machine, machine speeds, material specification, conditions of working, tools provided and allowances made, should appear, in addition to a detailed description of the operation itself, and should be based on the work elements observed during study. In some cases a drawing or diagram may help.

Where synthetic data has been developed, it is generally sufficient to draw up one specification for the whole range of time values covered. Subsequent standards issued from it can then be qualified by reference to this master. In those cases where a special form is used to compute values, this can sometimes serve as part of the specification, if it indicates the method of performing the work.

It is a mistake to issue too many copies of work specifications because of the difficulty of recalling them for amendment when there are changes. Where it is practical, it is best restricted to one copy which is retained in the work study department. At the most, there should be no more than two further copies in existence—one each for the Works or Production Manager and the Departmental Foreman. It must be made clear at the outset, however, that work specifications are available at any time for inspection by the Trade Union representatives.

To demonstrate the procedure more clearly, a suggested form of work specification for spraying paint based on the example in Chapter 3 now follows. It is given in four steps—

(*a*) Issue of standard times.
(*b*) Equipment used.
(*c*) Data specifications.
(*d*) Calculating standard times.

These are now explained in more detail.

(*a*) *Issue of standard times*

"Values for work operations will be issued by the Work Study Department in the form of standard times in minutes per 100 pieces. These will have been built up from synthetic data obtained by work study methods" (this could be elaborated upon if necessary), "and will contain reasonable allowances for contingencies, personal needs and rest.

"The parts of the operation used to build up the complete time values will also have standard times and they will be accompanied by a work specification. This will make it quite clear how much work is covered by the time given.

"Except where they are provisional or where there has been a genuine mistake, times will not be altered once they have been issued unless changes are made to the method of working, conditions, or the equipment. If this happens, the work will be re-studied and new times obtained.

"A standard time is the time during which an operator of average ability is able to perform the work when working at a brisk pace or target speed and taking the full relaxation allowance. These standard times will be issued on the understanding that they apply to work of a standard of quality required by the management.

"Time values obtained from the synthetic will cover the whole of the paint spraying operations at present being carried out in the spray shop. In addition to the allowances already mentioned, adequate time will be included for obtaining work which is to be sprayed from the stores, disposing of any surplus packing paper in which they were wrapped, switching on and off the paint spray compressor and stove and booking quantities of work done on a time sheet."

(*b*) *Equipment*

"The operations covered in the specification are to be performed using the following equipment—

1 Water wash spray booth—fitted with a turn table (specified by make and model number).
2 Special wire mesh trays (specified by drawing number or make).
3 Special trolleys (specified by drawing number or make).

"The layout of the department is shown on the accompanying drawing" (enclose a plan of the workshop).

(c) Data and specifications

Here it would be necessary to tabulate the various parts of the synthetic, giving a full specification for each part. For example—

(i) "Work per dial.
This includes taking dials from a storage container singly, wiping them clean with a rag, placing them on a turntable in the spray booth, covering them with an even layer of paint with the spray gun and laying them aside to trays. It also allows the removal of grit which may occasionally settle from the atmosphere, and re-spraying the dial afterwards."
(Then would follow a table of standard times.)

(ii) "Work per tray.
This includes moving trolleys of empty trays adjacent to the spray booth and taking the trays singly from this as required, placing trays full of newly-sprayed dials on to the conveyor, moving empty trolleys to unloading position and unloading trays of stoved dials from the conveyor to them, and moving trolleys full of stored dials to the print room. It also allows for surplus paint to be occasionally wiped from trays by means of a cloth dipped in paint solvent." (Then would follow the standard time value.)

(iii) "Work per charge of spray gun.
This includes charging the gun with paint from a small can filled from a drum and mixing the paint where necessary. It also allows for getting supplies of paint and paint solvent from stores and occasionally cleaning the spray gun by immersing it in solvent.

(iv) "Frequencies of work per tray and work per charge of spray gun.
These tables of frequencies represent the number of times that these amounts of work need to be performed per part sprayed. They have been obtained by study of the operation under actual working conditions."

(It should be noted that these would only form part of a whole series of such blocks of data covering the spraying of different articles with different paints.)

Final presentation of data

(d) Calculating standard times

This would give complete instructions of the method of obtaining standard times per operation. For example it might read—

"To complete a standard time:

(i) Take the work per dial for the appropriate size of dial.
(ii) Take the work per tray and multiply it by the frequency per tray for the appropriate size of dial taken from the table.
(iii) Calculate the work per charge of spray gun and multiply it by the frequency per charge of spray gun for the appropriate size of dial taken from the table.
(iv) Add these three values together and round off the result to the nearest whole number, the nearest five, or the nearest ten which is within $\pm 2\frac{1}{2}$ per cent of this addition."

This explanation could be followed by quoting an example similar to that given in Chapter 7.

Work specifications for single operations of a unique character would be similar to this except that they would be somewhat simpler in form.

Standard times for complete operations are usually issued to the shop floor by means of an *operation layout* (see Chapter 19) or a *job card* according to the type of production control system which is in operation. Where this is done, some reference should be made to the originating work specification, and the work specification itself should be referenced to the set-up. This enables any time value to be traced back if necessary to its originating time studies.

Storage and referencing of standards data

Great care must be exercised in the completion of synthetic data. No element should be classified as either constant or variable until sufficient supporting data has been collected. In some instances, either a basis for variation defies solution, or what appear to be constant elements vary for no apparent reason. This is undoubtedly due to insufficient evidence or illogical assumptions. Sometimes synthetics have to be partial in character, the unknown variable elements being individually studied until there are enough of them to establish a reason for their variation, when they can be built into the synthetic as a permanent feature.

Variable elements can often be interpolated from a graph, but they should never be extrapolated. For example, considering the operation of spraying dials shown on the *graph* Fig 19, times could be interpolated between sizes ranging from 2 in. to 15 in. dia., but not beyond. This would need more study and would extend the scope of

176

the synthetic. Work should always proceed in this way, and the situation should never be allowed to develop where each new time is individually studied and the synthetic allowed to deteriorate into disuse.

It is important that the elements used to form a synthetic are described in a standard manner throughout the organization. This will allow each to be classified, and if the classifications are cross-referenced on a card index or other convenient filing system, it will also ensure that none of these are studied more than once. Whenever extension becomes necessary in an area where synthetics already exist, the new data should be built into it so that the principle of compiling standard times by synthetic is preserved at all times.

Time values for complete operations computed from synthetic data should be carefully registered and filed in such a way that they can, if necessary, be traced back quickly to their originating synthetic and from this to the actual studies from which the synthetic was developed.

One way of doing this is to use the referencing system described in Chapter 18, combined with the use of special *computation forms*. These forms not only aid referencing, but make it possible to compute the standard times more quickly.

Use of special computation forms

An example of a special *computation form* for a machining and setting synthetic for capstan lathes is shown in Fig 41. This is shown completed for a brass casting used as a clock case. By using the methods described in Chapters 3 to 7 constant values for loading and unloading pieces to the machine were established. Machining times were obtained by process time study under optimum conditions of machining, constant values per unit length for various types of machining cut and various diameters being substantiated by plotting graphically.

The constant values were printed on the forms to save time in looking them up from tables. Machining values were tabulated for quick reference, values being rounded off to $\pm 2\frac{1}{2}$ per cent.

The setting synthetic was established by evaluating the average time it took to set-up and dismantle tools from the machine, standard times for these being established from the frequencies per set-up when such tools were used.

Standard times for machining and setting-up were used for a bonus incentive scheme for setter operators. Since the setting-up time values were used far less frequently than machining values, these were found to be sufficiently accurate if they were rounded off to within ± 10 per cent.

177

STANDARD TIME COMPUTATION SHEET—
CAPSTAN LATHES

Description _Case_
Operation· _Machine and polish_
Material _Brass_

Serial number _101_
Set up number _10_
Drawing number _532_

Part 1 — Machining			Std. mins per 100 occs	Occs per piece	Std. mins per 100 pieces
LOAD AND UNLOAD PIECES TO MACHINE	Spring collet		10		
	Screwed mandrel		70		
	Plain mandrel		50	1	50
	3 jaw chuck		20		
	Angle plate		30		

	Operation	Dia	Length			
MACHINE ON TURRET	Turn c/dia body	4"	1½"	40	1	40

	Operation	Dia	Length			
MACHINE ON CROSS SLIDE	Face c/all length	4½"	3/32"	10	1	10

	Operation	Dia	Length			
POLISH WITH EMERY CLOTH	Polish outside dia	4"	1½"	30	1	30
	Polish end	4"	3/32"	10	1	10

Machining time (standard minutes per 100 pieces)	140

Part 2—Setting		Standard minutes each	Quantity	Total standard minutes
FIT AND ADJUST TOOLS	Turret	4	1	4
	Cross slide	5	1	5
SET AND CHECK PIECE	Inspection department	6	1	6
	Setting time (standard minutes)			·15

Fig 41 Special computation form used for synthesizing standard
times for machining and setting on capstan lathes

Accuracy of standard times

The fine work measurement techniques described in Chapters 3 to 7 should develop standard times to within limits of $\pm 2\frac{1}{2}$ per cent. Time values obtained by using broad work measurement, although not so accurate, will in the majority of cases be satisfactory as a basis for bonus incentive, particularly if one of the differential or stabilized output incentives described in Chapter 14 is used.

Before a work measurement programme is commenced, however, some consideration should be given to the accuracy really necessary, so that work measurement can be as simple as possible, and the number of standard times kept at the minimum.

To understand the significance of this, consider the following example—

Write down 10 numbers ranging from 1 to 100 at random and add them together. Now round each of the original numbers to the nearest whole 10 digits. Add this second set of numbers together and compare it with the sum of the first set. In the majority of cases, the two totals will be surprisingly close to each other.

To explain this a little more clearly, take the following random numbers—

Random numbers	*Rounded off to nearest 10*
9	10
15	20
23	20
74	70
81	80
10	10
6	10
58	60
47	50
93	90
Totals 416	420

The difference between the two totals is less than 2 per cent.

An interesting conclusion can be drawn from this. Providing there are sufficient numbers in the set, the sum of any of 100 numbers (i.e. from 1 to 100 in units of one) will not, in the majority of cases, be very different from a sum made up from only 10 numbers (i.e. from 10 to 100 in units of 10). Which means that the number of values can be reduced to one-tenth without significant loss of accuracy.

Applying this to an actual situation, let the numbers 1 to 100 be standard times. If it could be guaranteed that operators would work on no less than 10 different jobs in a week, the number of time values could be considerably reduced by rounding off to wider limits, without greatly affecting the total weekly standard time. This means that both work measurement and the subsequent administration of the bonus system could be greatly simplified.

The table Fig 42 has been constructed from values obtained by statistical theory. To use this table, first determine the minimum number of different jobs which will be worked on during the bonus period (usually a week). Next, decide what limit of accuracy is required for the final credits of standard time within this period

Minimum number of different standard times	+ or − Guaranteed accuracy of individual standard times									
	5%	10%	15%	20%	25%	30%	35%	40%	45%	50%
1	5	10	15	20	25	30	35	40	45	50
2	4	7½	11	14½	18	21½	25	28½	32	35½
3	3	6	9	12	14½	17½	20½	23½	26	29
4	2½	5	7½	10	12½	15	17½	20	22½	25
5	2½	4½	7	9	11½	13½	16	18	20½	22½
6	2½	4½	6½	8½	10½	12½	14½	16½	18½	20½
7	2	4	6	8	9½	11½	13½	15½	17½	19
8	2	4	5½	7½	9	11	12½	14½	16	18
9	2	3½	5	7	8½	10	12	13½	15	17
10	2	3½	5	6½	8	9½	11½	13	14½	16
11	2	3½	5	6½	8	9½	11	12½	14	15½
12	1½	3	4½	6	7½	9	10½	12	13½	14½
13	1½	3	4½	6	7	8½	10	11½	12½	14
14	1½	3	4½	5½	7	8½	8½	11	12½	13½
15	1½	3	4	5½	6½	8	9½	10½	12	13
16	1½	2½	4	5	6½	7½	9	10	11½	12½
17	1½	2½	4	5	6½	7½	8½	10	11	12½
18	1½	2½	4	5	6	7½	8½	9½	11	12
19	1½	2½	3½	5	6	7	8½	9	10½	11½
20	1½	2½	3½	4½	6	7	8	9	10½	11½

Fig. 42 Table of expected accuracies (95% probability) of total standard times expressed as + or − percentages when different numbers of individual standard times within + or − of the percentages at the head of the column are added together

(usually this is $\pm 2\frac{1}{2}$ per cent, but wider limits could be considered, particularly if a stabilized output incentive is to be used). The accuracy to which standard times can be rounded off will then be shown at the head of the appropriate column.

For example, consider a case where there are at least 16 different jobs being worked on by an operator during a week. Using the table, it will be seen that individual values can be rounded off to limits of ± 10 per cent to allow an accuracy in the total of $\pm 2\frac{1}{2}$ per cent.

If time values which are to be used for repetitive and semi-repetitive work are expressed in the following units, then the numbers being used will tend to vary between the ranges of 50 to 1,000—

(*a*) Standard minutes per 1,000 for operations of half a minute or less.

(*b*) Standard minutes per 100 for operations taking more than half a minute.

The actual numbers of standard time values necessary within these ranges when rounding off to different accuracies are shown in the table—Fig 43.

To explain the use of this table, apply it to the example just quoted where it can be guaranteed that an operator will work on 16 different jobs in the week, and the individual values can be rounded off to within ± 10 per cent. Then, even if the full range of 50 to 1,000 is used, the total number of values will only be 25.

Reducing it to a procedure, this can be set down as follows—

(*a*) Determine the absolute minimum number of different jobs that any operator (or team of operators where a group bonus is being considered) should perform during the bonus period.

(*b*) Consult the table for the appropriate "rounding off" percentage.

(*c*) Write down the actual numbers which will be necessary for this "rounding off" percentage using the above table for this.

(*d*) If there are values already in existence, use both sets of values (i.e. the existing ones and the "rounded off" ones) in parallel for a period, to prove that the new total standard times are still within acceptable limits of the existing totals.

Whatever eventual "rounding off" percentage is decided upon, it is quite unnecessary to use a figure of less than $\pm 2\frac{1}{2}$ per cent, i.e. a maximum of 101 values in a full range of from 50 to 1,000.

An alternative method of reducing the number of standard times is to plot actual standard times in existence by their actual frequency of occurrence over a period. This will tend to give a curve with a

Final presentation of data

Range of values		Using the following percentage accuracies as a basis, round off to the nearest figures given below					
Over	To	$2\frac{1}{2}\%$	5%	10%	15%	20%	25%
50	100	2	5	10	10	20	20
100	200	5	10	20	20	20	50
200	300	10	20	20	50	50	100
300	400	10	20	50	50	100	100
400	500	10	20	50	100	100	200
500	600	20	50	100	100	100	200
600	700	20	50	100	100	200	200
700	800	20	50	100	200	200	200
800	900	20	50	100	200	200	200
900	1,000	20	50	100	200	200	200
Total number of values in the set		101	46	25	20	15	10

When rounding off to nearest	Round up when	Examples
2	Last digit is odd	71 is 72 but 72 is 72, etc.
5	Last digit is $2\frac{1}{2}$ or more	72 is 70 but $72\frac{1}{2}$ is 75, etc.
10	Last digit is 5 or more	74 is 70 but 75 is 80, etc.
20	Last but one digit is odd	69 is 60 but 70 is 80, etc.
50	Last 2 digits are 25 or more	124 is 100 but 125 is 150, etc.
100	Last 2 digits are 50 or more	149 is 100 but 150 is 200, etc.
200	Hundreds are odd	499 is 400 but 500 is 600, etc.

Fig 43 Guide to rounding off for standard times when working to a specified accuracy limit

series of peaks, the peaks representing modal averages at various levels. These modal averages can then be used as standards for values which lie evenly distributed either side.

Register of time values
An independent register of time values should be maintained in the work study department to provide reference back from operation

182

layouts or job cards to synthetic data and work specifications. The form of most of these can be very simple—being merely a standard stationery book ruled with columns for reference numbers, set-up numbers, brief descriptions and dates. It is particularly easy if *standard computation forms* are used which are serial numbered. The serial number will locate the form itself, from which it should be possible to trace the set-up number. This should locate both the work specification and its original set-up. From then on the data can further be traced back to its originating studies if necessary.

The effect of learning and interruption

Anyone doing an operation for the first time will tend to be clumsy and slow. But human beings soon learn by their mistakes, and gradually the speed of performance will rise with practice. A typical learning curve is shown below when starting from no experience.

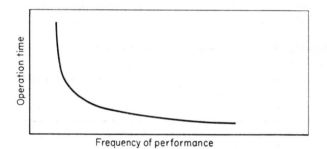

Fig 44 Typical learning curve

This is why learning allowances should be given in the initial stages of introducing bonus incentive (see Chapter 14). Learning curves, however, may sometimes progress in this way for months and sometimes years. This has the effect of making the standard times gradually become "looser". Because of this, it is wise to introduce a bonus incentive based on work measurement by agreement with the workmen via the appropriate Trade Union or other recognized body for a limited period of (say) three years, after which the work measurement, the method of operation and the rates of pay can be mutually revised and negotiated. Motion study followed by guided training can also be beneficial in such cases.

Continued interruption also has an effect on learning and rhythm as shown in Fig 45, each interruption causing a "saw-tooth" in the graph. This can mean that if batches of work become smaller after a period, the time values apparently become "tight". A means of

Final presentation of data

overcoming this is to publish values as a fixed "make-ready" time, plus a time per operation thereafter. For example, a standard time might be issued as 10 standard minutes per batch make ready time,

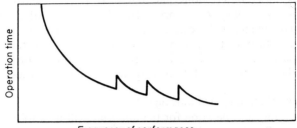

Frequency of performance

Fig 45 Graph showing the effect of interruptions

and 200 standard minutes per 100 pieces produced instead of (say) 210 standard minutes per 100 pieces reckoned on an average batch size of 100, etc.

Estimating the time necessary for work measurement
It is very difficult to estimate the time necessary to measure a sufficient range of work values to allow introduction of a bonus

Type of work	Time
Machine shop with capstan lathes; simple milling and grinding machine operations; and drilling and tapping using jigs and fixtures. The machining was carried out on small brass and steel components. Average batch size 1,000. Number of operators 50.	9 months
Dyeing processes in textile works. Included preparation of fabrics prior to loading them to the dyeing machines, process dyeing, unloading from the machines and spin-drying, and finally oven-drying. There were four types of dyeing machine and about ten different dyeing processes. Number of operators 31.	4 months
Tool room, using broad work measurement. Precision machining of all types was being carried out as well as fitting processes. Approximately 50 craftsmen involved.	15 months

184

incentive payment system, but some idea can perhaps be obtained from actual times taken from practice as set out on the preceding page. They were obtained with one fully-trained senior work study analyst with at least five years' practical experience and one senior assistant. They include the development of a full synthetic.

12

Method study

The human mind is a restless creation. Expose it to a problem and it will immediately grapple with it. And the more ingenious the mind, the more impressive will be the solution.

A labourer confronted with the simple task of moving earth a short distance with a shovel, will sooner or later develop a better way which will ease his burden or allow him to complete the job more quickly—even it is only using a truck or a wheel-barrow to help him. The same thinking processes motivate the engineer who designs a highly complex machine to increase productivity. The only difference is in the degree of sophistication. They have both examined alternative ways of doing the work, and by their reasoning arrived at an improved method. This is all that method study is—the examination of work performance and its subsequent rearrangement to produce the same result with less effort.

The scope of the technique is enormous; from a simple change of work pattern by moving a few benches or containers, to the design of a highly complex special purpose machine. But the solution of any problem must not be dictated by the impressiveness of the change. Very often, substantial cost reduction can be made by simple means with modest capital outlay, and although the design of special plant is sometimes the only ultimate solution, prior consideration must be given to the financial implications before launching any project, particularly a large undertaking. During investigation, it may be that one or two small operations are so obviously in need of improvement that there is temptation to tackle these first. This temptation should be resisted. It is far better to examine the overall function of the department and the feasibility of any possible improvements on this basis before turning attention to any, more detailed work. In fact, the application should be considered by degree—in the first instance on a broad basis, progressing later on to more detailed study. But the final decision as to which technique to apply, in which order —or whether any should be applied at all—must be decided by testing on an economical basis as explained in Chapter 18.

To practise method study one need not be a technician. On the contrary, a non-technical mind is often better equipped to grasp an overall situation and apply logical reasoning unhampered by any preconceived ideas. Nevertheless, technical knowledge cannot be ignored. It is often very important when considering highly specialized processes. On these occasions, management should encourage work study and technical staff to work closely together to encourage a harmonious merger between these two functions.

Method study should not only be applied after plant installation and workshop layout. The proper approach is to combine the work study team with process planning, factory planning and other related technical staff so that the best possible arrangement of work is carried out *at the planning stage*, thus lessening the need for costly modification later on.

Method study procedures

The very act of exposing methods in unconcealed detail will stimulate the mind; the more orderly the representation the greater will be the understanding and the greater will be the flow of ideas—particularly if there has been some specific training to direct thought towards this end. When carrying out method study, therefore, it is important to spend some time and care in both collecting and recording the information—not only to ensure accurate reproduction —but also to present the facts clearly so that they can be readily understood.

The procedure can be considered in five phases, a *preparation period* when the ground is prepared for the investigation; a *collection period*, when the information is collected from the activity centre by direct observation; a *presentation period*, during which recorded data is collated into a form more convenient for analysis; an *analysis period*, when the data is subjected to critical appraisal and rearrangement to simplify the procedures and bring about economies; and finally a *consolidation period*, when the improved methods are installed and consolidated.

Preparation techniques

Before any investigations take place, it is most important that preliminary discussions are held with supervision and representatives of the operators who will be affected by the study, in order that everyone is quite clear as to its nature and intention. This procedure is explained in more detail in Chapter 1.

Advice should also be sought from supervision on the selection of the most suitable operators for initial study (reference Chapter 3), and specialist staff consulted on any technical processes involved that are not completely understood.

By this time, some better understanding of the function of the area of operation will have been gained to allow a rough programme of work to be drawn up. This should indicate the limits and scale of the investigation and be formulated to collect the very minimum of information necessary for the intended improvements. It should also include details of the procurement of any special equipment used for the collection of data.

A drawing or sketch of the area is also an essential preliminary. In some cases, there may be architects' or planning engineers' layouts already in being, or interorganizational services may exist which can prepare these. But in every case, the critical dimensions must be checked on site to ensure accuracy. For it so very often happens that modifications to buildings, workshops, and office layouts are made without drawings being brought into line.

By whatever means the drawing is produced, however, it should be fairly accurately drawn to scale, and contain sufficient detail to clarify conditions as they appear in the area to be investigated.

Collection techniques

Method study is primarily concerned with movement—movement of a human being, a document, or a piece of equipment—and the events which happen during this progression.

If a cycle of operation takes many hours to complete, descriptions of the individual events must necessarily be broad, otherwise the amount of information available would be so excessive that it would be difficult to analyse. On the other hand, short cycle operations need to be broken down into very fine detail to be of any value. The means of collecting data must therefore be adapted to the purpose and intensity of the observation, different techniques being used for broader investigations than those for more detailed study over smaller intervals of time. They can be more conveniently described under the following main headings—

    (*a*) Recording the movement of equipment or material.

    (*b*) Recording clerical procedures.

    (*c*) Recording the work done by an operator on a broad basis.

    (*d*) Recording the work done by an operator on a more detailed basis.

(*e*) Recording the movement of operators with respect to machines or other operators working as a team.

(*f*) Recording information for micro-motion study.

(*g*) Recording the path of movement to improve workplace layout.

(*h*) Recording material wastage.

These are now explained in more detail.

(*a*) *Recording the movement of equipment or material*

Analysis of the movement of material, equipment or manufactured items can sometimes reveal excessively long distances travelled, backtracking movements, duplication of effort and the performance of unnecessary operations. These faults can often be alleviated by improving workshop layout, providing better storage and mechanical handling facilities, or altering the character and sequence of the way in which the work is performed.

In principle, the recording technique is simple. All that is needed is a series of *work study observation sheets* (Fig 9) clipped to a board similar to that used for time study, except that there will be no stop watch attached. Information taken down must conform to the scale of the investigation decided during the preliminary preparation stage. Lengthy operations need broader statements—at least in the first instance. Records in greater detail should be confined to the more intense study necessary to unravel the causes of local blockage in work flow.

Within the limits of the investigation, the path of movement of the material, parts, assemblies or equipment should be meticulously followed from the source of their travel to their ultimate destinations, by questioning all the operators involved and recording only what is seen to take place on the broad lines described below (the comprehensiveness of the information gathered will depend on the intensity of the studies being carried out)—

(i) A description of all the operations in terms such as "turn on lathe", "mill flats", "drill four holes", "fettle on grindstone", etc. Later, a more detailed investigation may need descriptions which subdivide these further and could include operational movements, inspections and delays in addition.

(ii) A description of all major movements which occur in connection with the work such as "transport from stores to assembly department", "load by crane to lorry", etc. Again, these may be subject to subdivision later on if deeper investigations follow. Where there are complicated movements back and forth, it may be necessary to draw a rough sketch on site, or

189

specify the limits and position of the movements from which a scale diagram can be constructed.

(iii) A description of all inspection processes defining its scope, i.e. whether it is 100 per cent or less.

(iv) A description of all counting, weighing or other quantity measurement operations.

(v) A note of all delays occurring not due to storage.

(vi) Details of storage operations together with the location of the stores.

(vii) Measurements of distances travelled during major movements. These may be obtainable from the site plan if points of travel are noted as they occur.

(viii) Descriptions, drawings or samples of all the individual parts, sub-assemblies and assemblies covered by the investigation.

(ix) The number of operators engaged on the performance of each event recorded, together with details of their names and works numbers (if any).

(x) Quantities of parts or other material involved.

(xi) Quantities of parts and material rejected or wasted due to the nature of the process or faults in manufacture, and the routes followed by these.

(xii) Approximate times taken—where this is included in the original terms of reference, or where it has been cleared by management and shop floor representation as being permissible.

(xiii) Records of any equipment used, machine operating characteristics and any other details considered relevant to the investigation.

No more detail should be recorded than is absolutely necessary, and no effort should be spared to ensure completeness of data, even if it means rechecking information previously recorded, or taking further studies.

(b) Recording clerical procedures

This is used to record existing clerical routines with a view to simplifying them, reducing clerical cost or increasing the effectiveness of the information dispensed.

The recording technique again follows a similar method to that used when following the movement of material or equipment, writing down what is seen to take place on observation forms clipped to a suitable board. The progress of individual documents can be followed by questioning all the persons involved in their preparation, or alternatively the activities of each person can be separately studied. The first method tends to cause less disturbance with the function of the department, but the second method ensures that no "hidden" document is missed.

Often, it is possible to reduce investigation time by persuading office workers temporarily to record their own activities over a short period, and then to combine this with a lesser amount of direct observation.

The ultimate aim is to collect information on the following lines, careful note being taken of the number of copies of each document and the route that each copy takes—

(i) A description of all the clerical operations in terms such as "receive goods inwards note from stores, enter quantity and update stock control card", "enter details of bonus payment from daily work sheet to payroll form", etc. Later, it may be decided to break some of the operations down to further detail as the investigation proceeds.

(ii) Details of the movement of documents as, for example, "deliver from purchasing department to accounts department", etc.

(iii) A description of any checking procedures carried out on any of the documents.

(iv) Notes of any delays in progress of a document together with the reasons for these.

(v) Details of any filing operations.

(vi) Details of distances travelled and routes taken by documents, particularly inter-office movements.

(vii) Samples of all forms and documents used, pinning these to the observation sheets.

(viii) Details of the functional responsibilities of the personnel together with their names and works numbers. This, for example, could be "J. Smith, stock control clerk, production control office", etc.

(ix) Details of any office equipment or machinery used together with their operating characteristics, and any other details relevant to the investigation.

(c) Recording the work done by an operator on a broad basis

The procedure for recording work of this nature is similar to that used for investigating the movement of equipment or material. The only difference that the operator is followed instead of the material. It is again important to decide the depth and scope of the investigation and confine it within limits to avoid over-complication.

Examples of its use are in studying the activities of operators engaged on fairly lengthy tasks such as occur in maintenance departments, general offices, assembly shops, etc.

(*d*) *Recording the work done by an operator on a more detailed basis*

This is used for fairly short cycle work generally performed at a workplace, where operations can be broken down into recognizable elements of work in a similar way to that used for time study, except that the movements of the hands and other limbs are considered separately and in relation to one another. The equipment necessary is a clip board and a series of *work study observation sheets* (Fig 9). The operation should be carefully watched over several cycles so that the observer can familiarize himself with the purpose and rhythm of the work. A convenient point is then chosen at which the observation can commence. Any recognizable point will suffice, but it should be preferably when the hands are in unison. This will not necessarily be at the start of the operation. Each main column on the observation form is headed "left hand" and "right hand", and the movements of each hand entered, care being taken to record them in correct sequence with simultaneous movements opposite each other on the form.

Because of the speed of the operation, several cycles may have to be observed before the rhythm of the whole can be written down, the procedure being to record as many sequential movements as possible cycle by cycle until complete. The operation should then be watched over a further few cycles as a final check for accuracy.

It will be of advantage to subsequent analysis if studies are taken of the same operation on more than one operator.

Descriptions of elements should be of the following magnitude—

      pick up part from container.
      place part in fixture.
      operate machine.
      twist to unscrew.
      hold part in hand.
      pick up gauge.
      inspect part for length.
      put down gauge.
      place part aside on conveyor, etc.

(*e*) *Recording the movement of operators with respect to machines or other operators working as a team*

When investigating activities of this nature it is necessary to record the work content as well as the method of working of each member of the team, and the cyclic pattern of any machines used. Sometimes, data can be recorded by time study methods (reference Chapter 3), or by using one of the broad work measurement techniques described in Chapter 8. If the team workers are in close proximity to each other, it may be possible for one observer to record the complete procedures by using a specially constructed study board such as is described in Chapter 8, and which is designed to hold up to four observation forms at a time. Another method is to study each operator in turn to obtain the time values, finally observing the complete procedure yet again to note the point of interaction of the various tasks. Similar methods might be used to study the work performance of operators feeding a machine.

Quite often, however, the work is so complex and confusing that the only practical way possible is apparently to use one observer to study each operator and machine simultaneously. But this is often difficult to arrange, particularly if the work study staff is limited.

An alternative method is by memotion study. This involves the use of a special ciné camera which is adapted to run at speeds ranging from two frames per second to one frame in four seconds. If this is set up to film the complete area under review, a camera with 100 ft film capacity could be arranged to operate continuously over a period of four hours taking one frame every four seconds. A clock fitted with a seconds hand should be placed in view to act as a counter. This method, however, needs non-standard equipment which is difficult to obtain, and may have to be specially made by modifying ordinary cameras and projectors to suit. It is then possible to analyse the operation by projecting the film several times at a speed convenient for analysis.

(f) Recording information for micro-motion study
It is largely impracticable to attempt to analyse some of the very fast short cycle repetitive operations into detailed movements by the naked eye alone. The problem can be overcome, however, by using a ciné camera to record the process and to project the film a frame at a time afterwards.

Any good quality sub-standard equipment is suitable for the purpose although the best results will be obtained if the camera has certain refinements.

16 mm film gauge is preferable to 8 mm, as the projected image will be clearer. The capacity and drive of the camera should be such as to allow at least four minutes uninterrupted filming at 16 frames per second. A variable speed drive allowing filming to take place at 8, 16, 24 and 64 frames per second is a useful refinement to observe very fast or very coarse movements. A zoom lens, or a turret

head with alternative lenses is essential so that shots can be taken at close range or where the camera cannot be placed too near the subject. The ideal type is perhaps a through-the-lens reflex camera with a built-in exposure meter and footage indicator. With this there is less chance of failure since the whole of the image seen in the view-finder will be recorded on the film. A tripod fitted with a pan and tilt head will facilitate camera operation and ensure a rock-steady image on projection.

Before filming, ensure that the workplace has adequate lights, using photoflood lamps to augment the normal illumination if necessary. A counter to register the speed must be placed in the field of vision. This should contain three dials, with pointers which sweep them radially. The main dial should be graduated in one hundreds and the speed of the pointer 20 rpm, each graduation then registering one two-thousands of a minute, which will then correspond to the original "wink" of Frank Gilbreth. The other two pointers revolve at successively slower speeds to record the number of revolutions of the pointer on the main dial.

The camera is loaded with high-speed monochrome film and set up in a convenient position so that the whole of the operator as well as the operation is clearly visible in the camera view-finder. The first few frames are then exposed to a blackboard on which has been chalked the study number, operator's name and brief description of the process. It is then necessary to ensure that the operator is as relaxed as possible and observe the operation being performed normally through the view-finder before operating the camera.

For very short cycle operations, film more than one cycle. Longer work cycles needing more than one shot should overlap to ensure a complete record. Throughout the filming the moving counter must be clearly visible.

Take observations on different operators if at all possible, as later comparison may tend to stimulate ideas for improvement. Label each shot with new details on the blackboard.

After processing the film, ensure that the first persons to see it projected include the operators observed.

Human activity was analysed by Frank Gilbreth into elemental movements which he called "therbligs". These are described later in this chapter in the section headed "simo charts".

Very experienced observers are capable of analysing movement patterns into therbligs on some operations by direct observation, watching several cycles of the work take place and successively recording these until the whole cycle is covered. A clip board and a series of *motion analysis sheets* (see Fig 51) are all that is necessary for this purpose, the therbligs being indicated by standard symbols.

194

(g) Recording the path of movement to improve workplace layout
Observations made of the performance of repetitive work sometimes reveal cumbersome and unnecessary movements. Improvement in layout of the workplace by the provision of better storage
and working facilities will often reduce or eliminate these and bring
about reductions in time cycle for very little capital cost.

Techniques for collecting the information are either by direct
observation or photographic methods.

Direct observation can sometimes be adequate, but is often
lengthy, especially if the movements are being performed at a fast
rate. One way of overcoming this is to observe the work being done
at a slower pace and record the position in space of the operator's
hands at various periods during the performance of the work, and
from this to draw a complete diagram of the paths traced out.
This method, however, is difficult to carry out in practice and a far
better way is by using chronocyclegraph apparatus.

For this, one needs a special device which will cause low voltage
lamps to flash suddenly into brilliance and die away more slowly
after being switched off. Provision should also be made to allow the
rate of flashing to be varied between 10 and 30 cycles per second.

If lights are then taped to each of an operator's wrists using a
thin wire in such a way that there is no interference with movement,
a still photograph sufficiently exposed to the area will record the
path traced out by each hand in a series of pear-shaped dots, the
tail-end of the "pears" indicating the direction of movement. The
faster the movements, the closer will the dots be together. If colour
photography is used, the different paths traced out by each hand
can be shown by attaching different coloured lights to each wrist.

(h) Recording material wastage
Material waste sometimes results in substantial monetary loss. It
could be due to any of the following reasons—

(i) Excessive work in progress or storage, causing loss through
obsolescence, damage or deterioration.
(ii) Lack of care and attention by operators or incorrectly operating
machines, causing faulty manufacture or extravagant material
useage.
(iii) Over- or under-production due to inadequate counting methods.

Although excessive work in progress can sometimes be alleviated
by improvements to workshop layout, the real cure for this and
excessive material stocks is better production and stock control.
Material losses due to operating faults, however, can very often be
reduced by the application of method study, although technical

advice may be needed in addition when dealing with specialized processes.

The movement of the material through the various operations should be recorded using collection techniques (*a*) described above. It will also be necessary to take a material audit at each point where there are losses, carefully recording the quantity of material used and the quantity of material wasted. This can be carried out in various ways according to the nature of the loss.

Where shapes are being cut from pieces of material as occurs in the textile, leather, engineering and other allied industries, or during such operations as presswork, etc., record the description, part numbers, sizes and quantities of every part produced segregated into acceptable and non-acceptable quality groups. The total quantity of the raw material used should also be recorded, together with any possible reason for the wastage.

Some machining processes cause unavoidable losses, such as "short ends" on centre lathes or automatic and semi-automatic turret lathes; feeder and sprue groove material in die-casting and plastic injection moulding; material discard in extrusion processes, etc. But these losses can sometimes become excessive due to inconsistent tool design, improvident machine setting or faulty tool operation, incorrect machine operation, etc. In these cases, details should be recorded of the method of working as well as the amount of the material used, and the amount of acceptable and non-acceptable parts produced.

Material loss can also occur due to tool wear. For example, an automatic turret lathe may be periodically stopped to allow the tools to be re-sharpened and re-set. There is a point in the process, however, at which tool wear causes the machine to produce reject work due to an oversize condition or too great a fall in the standard of finish. If parts are taken from the machine at regular intervals and examined, the point at which rejects commence can be found by plotting results on a graph. Preventative material wastage can then be operated by arranging to break down the machine at a point just prior to the time when reject work would start to be produced.

Other sources of material waste sometimes occur in plating processes. On these occasions excessive thicknesses of metal may be deposited or excessive scrap produced due to neglect in maintaining plating solutions or the design of the jig which forms the electrode. Here, the method study analyst can sometimes help by obtaining plating thicknesses from coulometer tests taken by technical personnel, and examining with them different methods of suspending plated parts to develop different jig designs to produce uniform minimum thicknesses of deposit within the specified limits.

Inadequate counting methods are best investigated by arranging for physical checks by weight, etc., and comparing results with the quantities normally recorded using existing practice.

Presentation techniques

Method study data is chiefly presented in the form of charts. These are of several types—

Process charts which are charts which do not have to be drawn to any scale, and use symbols to represent the activities, the lines linking these indicating their sequence.

Flow charts and string diagrams, etc., which are charts drawn to a distance scale with arrowed lines or coloured threads to show the progression of the work in space. In addition some of these use the same symbols as those on process charts to represent individual activities. These are known as *flow process charts.*

Multiple activity and simo charts which are charts drawn to an exact or approximate time-scale in histogram form.

Cyclegraphs and chronocyclegraphs which are photographic records showing the path of movement of an operator's hands during the performance of work.

Document or material audits and analyses which are usually represented in tabular form, graphs also being used to illustrate certain material characteristics and changes.

Process chart symbols

Several symbols are employed for *process charts.* As it is important not to sacrifice clarity for elaboration or brevity, it is recommended that in their construction the six shown on p. 198 are used.

Preparation of data for the construction of process charts

Study data recorded on *work study observation sheets* (Fig 9) should be entered to a *process chart sheet* (Fig 46). Study numbers allocated from a *study register* (Fig 11) are then entered to all observation and process chart sheets together with sequential numbers and relevant details of the department, process, product and operation.

Each process or element described on the *observation sheets* is allocated a reference letter, and the same reference letters are entered under the "Ref." column on the *process chart sheets,* descriptions of these being written neatly and legibly against each.

Activities are classified according to the *process chart* conventional symbols, and the symbols themselves drawn against each under the appropriate column. For broad method study it may be adequate to use only one or two of the symbols (i.e. the operation

197

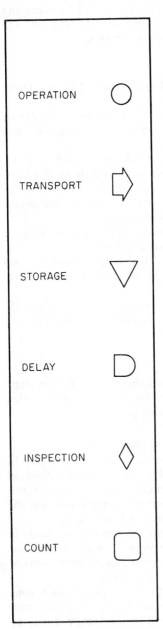

OPERATION — Represents the main steps in a procedure tending to change the shape, size or classification of a product, material or document.

TRANSPORT — Represents the movement of workers, material, equipment or documents from one place to another.

STORAGE — Represents a controlled storage operation during which material is in a store or a document is in a file, or an item is retained in a receptacle preparatory to further controlled use or transport.

DELAY — Represents an uncontrolled storage or delay in the sequence, when work is waiting between consecutive operations or events due to some fault in the process or procedure.

INSPECTION — Represents a recognized check for quality. This would include checking parts for accuracy or appearance, or checking correctness of data on a document, etc.

COUNT — Represents a check for quantity by physical count, weighing, or other quantity measuring methods.

PROCESS CHART SHEET

Department __Sales__	Study No. __85__
Section __Finished goods warehouse__	Sheet No. __1__ of __1__
Product __Various__	Date __17·8·71__
Operation or Process __Select goods__	Taken by __B. R. Ford__
__from finished stores, pack and__	Operator __J. Fox and S. Baines__
__load to despatch vehicle__	Plant details __Stores trolley__
Remarks __Average of estimated__	__type B3 and 3 ton__
__times at standard performance__	__despatch vehicle__
__entered.__	

Ref.	Description	Symbol	Dist. (feet)	Time (mins)	Quantity Ops.	Quantity Units	Time per unit
A	Count quantity of goods from warehouse shelves	☐	–	18	1	200	0·09
B	Place on stores trolley	◯	–	22	1	200	0·11
C	Wheel trolley to packing area	⇨	38	0·3	1	200	··
D	Pack goods on trolley into cartons and outers	◯	–	40	2	200	0·20
E	Address outer cartons	◯	–	10	2	200	0·05
F	Move goods by hand to despatch bay	◯	25	13	2	200	0·07
G	Wait for arrival of despatch vehicle	D	–	20	2	200	0·10
H	Load goods to despatch vehicle	◯	–	15	2	200	0·08

	Operation	Transport	Storage	Delay	Inspection	Count
SYMBOLS	◯	⇨	▽	D	◇	☐

Fig 46 Process chart sheet. This can be used to summarize methods study investigations prior to construction of process charts

symbol alone, or the operation symbol combined with either the inspection or transport symbol). Only use the full range of symbols when it is necessary to do so for the sake of clarity.

When right- and left-hand movements have been described, a process chart sheet can be completed for each hand, the sequence of the movements being indicated by allocating letter references for each hand progressively. For example, the first movement (by the left hand) could be described as AL and the corresponding movement at the same time performed by the right hand (which could be a delay if this is unoccupied) as AR, etc.

Distances travelled for the "transport" function should also be added where these are substantial. It is also of advantage in certain instances to record the standard time taken and enter this on the form. In those instances where more than one operator and more than one unit are involved, these details should also be entered under the appropriate columns.

The "time per unit" is calculated by multiplying the total time by the number of operators and dividing by the number of units. For example, if two men move 500 lb of material a certain distance in 20 minutes then the time per 100 lb (unit) would be

$$\frac{2 \times 20}{500} \times 100 = 8 \text{ man-minutes per 100 lb.}$$

After completion of the studies in this way, a return visit should be made to the area of operation and all details checked against actual conditions. Further studies should be taken if there is any discrepancy until absolute authenticity of the recorded data is assured.

Constructing process charts

Process charts should be constructed extremely neatly and in ink, preferably using stencils or drawing instruments as an aid. Coloured inks denoting the different symbols do often help clarity of presentation.

Construct a trial chart free hand in pencil first, from memory if possible. It is usually better to start from the end of the main operation sequence and work back to the beginning, checking this through on site to ensure accuracy before proceeding further.

Descriptions must be consistent and to the same scale, such items as "screw on nut and tighten with spanner" should not appear on the same chart as more detailed items as, for example, "procure screw", "assemble to tapped hole", "procure screwdriver", "tighten screw", and "aside screwdriver to bench", etc.

The depth and nature of an investigation determines the form of the chart. Simple charted statements may sufficiently explain broad investigation, whereas for more detailed study they will need to be more elaborate. When reviewing very complex operations, it may be necessary to show successive detail on several charts rather than attempting to include everything on one alone.

Examples of different techniques are given as follows—

(i) *Simple process charts* (Fig 47).

These are the simplest type and represent the sequence of events in the form of symbols joined together vertically. They can be constructed either by using one, two or the whole six symbols, combining some of these together if appropriate. Operation descriptions are printed at the side of each symbol. Distances travelled and measured time values can also be added if these will clarify presentation. Charts of this type can be constructed to show either the movement of men, material or documents.

(ii) *Multiple process charts* which record more than one activity or process (Fig 48).

Charting can be adapted to show the interaction of different sequences of events on each other. Examples include those which illustrate the relative movements of the right and left hands during the performance of work (Fig 48A): the sequence of activity which occurs during the manufacture of individual parts which are subsequently assembled together (Fig 48C): operations which start as one sequence and divide into others, such as occurs in office work when more than one copy of a document is produced, each following a different route (Fig 48B); or when products are classified by inspection into various types of recoverable rejects and scrap, and the different patterns of activity and routes for rectification and destruction need to be shown, some of these possibly rejoining the main flow later on.

As with *simple process charts*, this type can be constructed using one or more symbols up to the full six according to the amount of detail which requires presentation. Supplementary information such as distances travelled, times taken, etc., as well as descriptions of the work elements or operations can also be entered for added clarity.

(iii) Charts which represent movement to an approximate distance scale.

Some operations can be expressed more clearly by constructing a chart on an approximate scale of distance.

The *process chart* shown in Fig 49A represents a simple sequence of operations being performed by a worker who

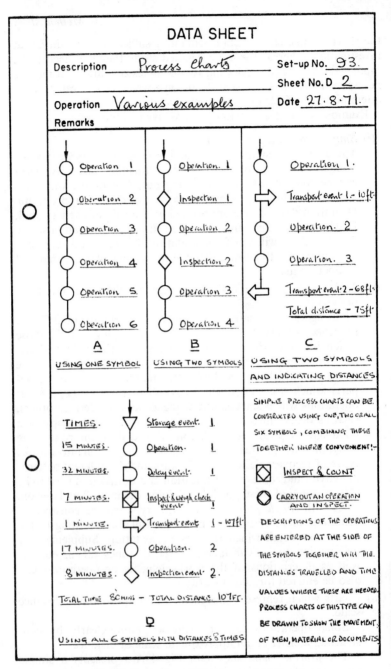

Fig 47 Examples of simple process charts

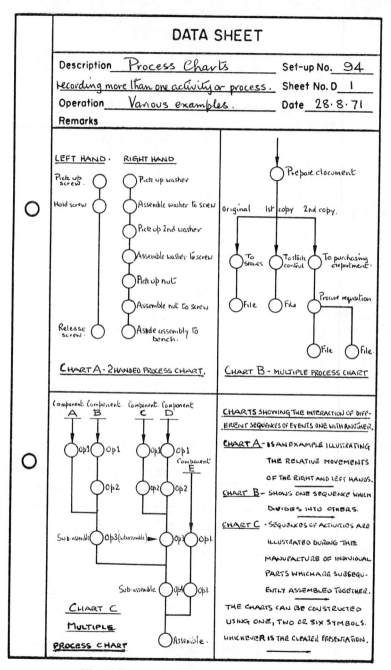

DATA SHEET

Description __Process Charts__   Set-up No. __94__

__Recording more than one activity or process.__   Sheet No. D __1__

Operation __Various examples.__   Date __28·8·71__

Remarks

LEFT HAND.   RIGHT HAND

Pick up screw ◯ — ◯ Pick up washer

Hold screw ◯ — ◯ Assemble washer to screw

◯ Pick up 2nd washer

◯ Assemble washer to screw

◯ Pick up nut

◯ Assemble nut to screw

Release screw ◯ — ◯ Aside assembly to bench.

__CHART A - 2 HANDED PROCESS CHART.__

◯ Prepare document

Original   1st copy   2nd copy.

◯ To Stores   ◯ To stock control   ◯ To purchasing department.

◯ File   ◯ File   Procure requisition

◯ File   ◯ File.

__CHART B - MULTIPLE PROCESS CHART__

Component A   Component B   Component C   Component D

◯ Op1   ◯ Op1   ◯ Op1   ◯ Op1   Component E

◯ Op2   ◯ Op2   ◯ Op2

Sub-assemble ◯ Op3 (subassemble) — ◯ Op3 ◯ Op1.

Sub-assemble ◯ Op4 ◯ Op1.

◯ Assemble.

__CHART C__

__MULTIPLE.__

__PROCESS CHART__

__CHARTS__ SHOWING THE INTERACTION OF DIFFERENT SEQUENCES OF EVENTS ONE WITH ANOTHER.

__CHART A__ - IS AN EXAMPLE ILLUSTRATING THE RELATIVE MOVEMENTS OF THE RIGHT AND LEFT HANDS.

__CHART B__ - SHOWS ONE SEQUENCE WHICH DIVIDES INTO OTHERS.

__CHART C__ - SEQUENCES OF ACTIVITIES ARE ILLUSTRATED DURING THE MANUFACTURE OF INDIVIDUAL PARTS WHICH ARE SUBSEQUENTLY ASSEMBLED TOGETHER.

THE CHARTS CAN BE CONSTRUCTED USING ONE, TWO OR SIX SYMBOLS. WHICHEVER IS THE CLEARER PRESENTATION.

Fig 48 Examples of multiple process charts

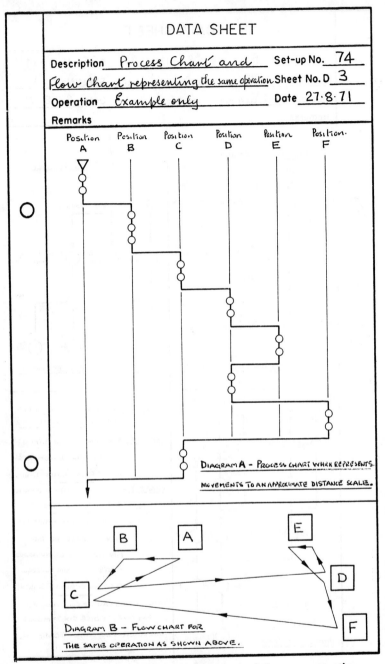

DATA SHEET

Description Process Chart and   Set-up No. 74
Flow Chart representing the same operation Sheet No. D 3
Operation Example only   Date 27·8·71
Remarks

Position   Position   Position   Position   Position   Position
   A   B   C   D   E   F

DIAGRAM A - PROCESS CHART WHICH REPRESENTS.
MOVEMENTS TO AN APPROXIMATE DISTANCE SCALE.

B   A   E

C   D

DIAGRAM B - FLOW CHART FOR   F
THE SAME OPERATION AS SHOWN ABOVE.

Fig 49 Diagram showing representations of the same operation
by process chart and flow chart. Diagram B is also referred to
as a flow diagram

moves his location in order to complete the job. These can be indicated by a shift in position on the chart to provide illustration of some of the backtracking movement involved.

The same process is drawn as a *flow diagram* in Fig 49B, the two diagrams together expressing the conditions extremely clearly.

Preparation of data for constructing flow charts and string diagrams
This is a similar procedure to that used for preparing data for *process charts*, except that the exact location of the points of movement will have to be clearly indicated on the *process chart sheets*.

Constructing flow charts
Flow charts are constructed simply by marking the points of movement on a site plan of the area, and joining these together in sequence, arrowing the lines to show the direction of travel. They can be constructed in plan view (Fig 50A); in plan and elevation; or on a perspective drawing (Fig 50B). Process chart symbols can also be added if these clarify the presentation (Fig 50A).

String diagrams
String diagrams are *flow diagrams* which show repeated movements back and forth but using threads instead of lines for the purpose. A site plan is pasted or pinned to a solid board and pins driven at the points of movement and at the corners of natural obstacles such as partitions, benches, machines, etc. Threads are then wound around the starting points of travel and stretched to finishing points in the way that the movement has taken place. The threads can be marked off in lengths to indicate distances travelled, different colours being used if there is more than one path of movement recorded on one chart.

Multiple activity charts
These show the relative activities of operators to a machine or process, or to other members of a team, on a time basis. They were discussed in Chapter 9, with further examples in Chapter 10.

Simultaneous motion charts (SIMO) chart symbols

The symbols used for the preparation and construction of simo charts are *therbligs*, which are the basic elements of motion of the human body originated by Frank F. Gilbreth. There are eighteen of these, each having an identifying symbol and colour as follows—

SEARCH (black)		Represents the effort of looking for a particular object in a group of objects.
FIND (grey)		This marks the end of the "search" therblig.
SELECT (light grey)		Usually this element follows "find" when the hand moves towards the object in order to pick it up.
GRASP (red)		Represents the process of securing an object by the hand.
HOLD (gold ochre)		Represents a delay in movement whilst grasping the object.
TRANSPORT LOADED (green)		Represents the movement of an article whilst being grasped.
POSITION (blue)		Represents the movement necessary to locate an article for further action after moving it from one place to another.
ASSEMBLE (violet)		Represents the action of assembling two or more parts together or fitting a tool to a part prior to using it.
USE (purple)		Represents the performance of work when using a tool.
DISASSEMBLE (light violet)		This is the reverse of "assemble", covering also the disconnection of a tool after use.
INSPECT (burnt ochre)		Represents the examination of a part for defects.

PREPOSITION (pale blue)		Represents the placing of an article or tool in such a position that it is ready for immediate use.
RELEASE LOAD (carmine red)		Represents the reverse of "grasp", when the hold by the hand or a tool is released.
TRANSPORT EMPTY (olive green)		Represents the movement of the arm in transferring the hand from one point to another after completing one operation and starting the next.
REST FOR OVERCOMING FATIGUE (orange)		Represents a necessary relaxation period.
UNAVOIDABLE DELAY (yellow ochre)		Represents an inactive period due to the nature of the method of operation and which is outside the operator's control.
AVOIDABLE DELAY (lemon yellow)		Represents an inactive period which can be controlled by the operator and is due either to inattention or lack of training.
PLAN (brown)		Represents a period of delay in movement when planning the next move.

Preparation of data for the construction of simo charts

A silent film projector fitted with a reversing mechanism and capable of projecting a single frame indefinitely without burning the film is necessary for this purpose. As previously mentioned, it is better to use 16 mm than 8 mm equipment for the sake of greater clarity of projected image.

Run the micro-motion study film through the projector several times at normal speed (16 frames per second). If more than one cycle has been recorded, check the relative operation times with the counter readings on the film and choose the most representative for analysis. Then note the most convenient starting point in the cycle.

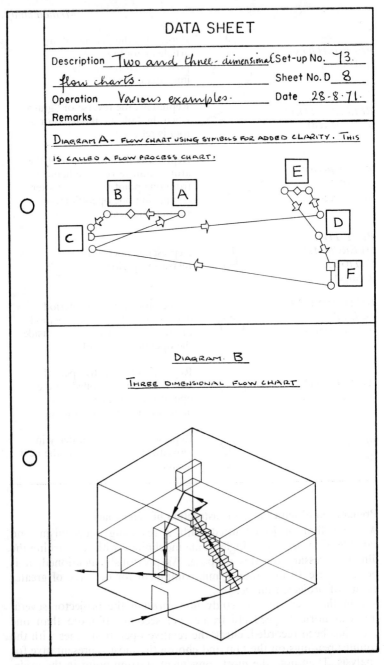

DATA SHEET

Description <u>Two and three-dimensional</u> Set-up No. <u>73.</u>
<u>flow charts.</u>   Sheet No. D <u>8</u>
Operation <u>Various examples.</u>   Date <u>28·8·71·</u>
Remarks

DIAGRAM A - FLOW CHART USING SYMBOLS FOR ADDED CLARITY. THIS
IS CALLED A FLOW PROCESS CHART.

DIAGRAM. B

THREE DIMENSIONAL FLOW CHART

Fig 50 Examples of two- and three-dimensional flow charts.
Diagram B is also referred to as a flow diagram

It should be easily recognizable—preferably when the operator's hands are working in unison. Invite the operators on whom the observations were taken to this preliminary showing. It will do a great deal to establish good relationships.

Detailed analysis is carried out by operating the projector a frame at a time until there is a change of therblig element. Both hands are considered at the same time during this analysis. The therblig symbol, brief description and counter reading are recorded on a *motion analysis sheet* as shown in Fig 51.

Constructing simo charts

Simo charts should be constructed from data recorded on the *motion analysis sheet* using 1 in. square graph paper subdivided into one-tenth divisions. The movements of each hand should be charted side by side, it being a further advantage if the movements of the arms, fingers and eyes are in separate columns laterally (see Fig 52). The use of the standard colours will aid interpretation.

The vertical scale of the chart is in "winks" as originally taken from the ciné film which photographed the wink counter readings.

Cyclegraphs and chronocyclegraphs

These are actual photographic records of the paths traced out by lights attached to operators' wrists as described earlier in this chapter. A *cyclegraph* is a photograph taken using a constant light source and a *chronocyclegraph* uses a flashing light. The latter has the added advantage of showing the direction and relative speeds of motion by the shape of the "dots" and the distance they are apart.

Document analyses

These are charts or tables which show the dissemination of information for existing documents to the various departments and executives.

Each document can be appraised under the following headings—

Title of form
Information entered
No. of copies
Frequency of completion
Method of completion
Originating authority
Receiving authorities
Final destination

MOTION ANALYSIS SHEET

Department	Purchasing Office	Study No.	85
Section	Filing	Set up No.	19
Product	Suppliers Records	Date	17·8·71
Operation	Check record with	Taken by	T.M.
delivery invoices		Operator	A.S
		Plant details	Manilla
Remarks First sheet only –		folders used as files	
continued in Sheet No 2.			

Time	Record-ing	PRESENT METHOD LEFT HAND Ther-blig	Description	PRESENT METHOD RIGHT HAND Ther-blig	Description	PROPOSED METHOD LEFT HAND Ther-blig	Description	PROPOSED METHOD RIGHT HAND Ther-blig	Description
0	164	∩	File	⌒					
12	176	∪							
15	179			∩	Paper				
30	194	∩							
32	196	∪							
38	202	∩	Regrasp						
58	222	#		#	Remove paper				
92	256	∪		∪					
94	258	△		∩	Paper				
105	269	∩	File	∪					
107	271	∪		∩	Pencil				
126	230	∩	Paper	∪					
128	232	△		9					
140	304			∪	Mark				
152	316			∪	Aside pencil				
156	320	∩		∩					
158	322	∪		∪	Pencil				
164	328	⌐O	Delay	∩	Paper				
170	334			∪					
174	338			9					

Fig 51 Motion analysis sheet. This is used to record the result of a cine film taken for motion study. It can also be used for direct observation using therbligs

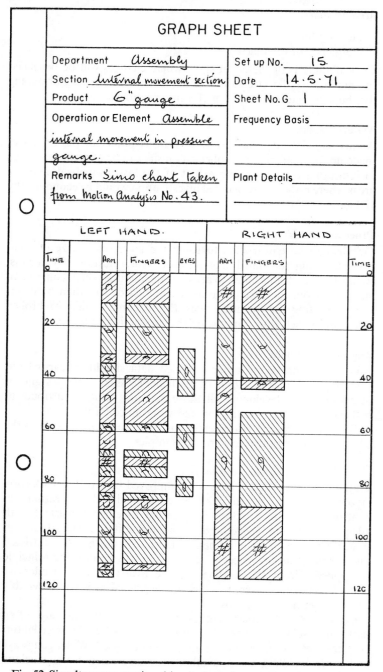

Fig 52 Simultaneous motion (simo) chart. In practice, the standard
colour codes would be used to denote therbligs

Time on file

Frequency of reference

Critical analysis of these can show where changes will be beneficial by eliminating the circulation of unnecessary data and arranging for the compilation of more useful information in its place.

Material audits

Audits showing material losses at various points in the processes made up from information collected on site can be of great assistance in indicating the reason and extent of the losses involved, and the areas where rather more detailed investigation would be economically viable.

Process analysis

Graphs can be constructed to show the increases in size of components being machined due to tool wear, plotting size against numbers of parts machined to arrive at the point at which component size is just outside acceptable limits. Standards can then be determined quoting the quantity when the machine should be broken down and re-set, thus minimizing the production of scrap.

Analysis techniques

Analysis of charts and other collected data should be carried out with the principles of eliminating all unnecessary activities and of promoting and developing all those activities which contribute to better overall control and improved working conditions.

To explain these in more detail, consideration is now given to the effect of applying them to various examples.

(a) Improvements to material flow

Material flow can nearly always be improved by a better work-shop layout. Less distances will need to be travelled if the operation and control points are as close as possible together consistent with adequate working space and manœuvrability. Material should flow, if possible, in straight lines, and although this is often restricted due to limitations in space, a better layout will result if the principle is applied within the confines of the available area. Under no circumstances should contra-flow be considered as it will inevitably cause congestion and chaos.

Conveyors and other mechanical handling devices are often useful in disciplining work flow and tend to reduce indirect labour cost, but they should never be indiscriminately used. Wider gangways and more handling space is often necessary when using trucking methods and this will mean more floor space. Very careful consideration must be given to capital costs and increased running costs set against the financial benefits, for often it will be found that simpler and less costly methods are just as effective.

Different storage methods sometimes improve material flow. The use of those standard size unit containers which will store material both on the shop floor and in the stores will often considerably reduce handling times and simplify stock control and production control.

Work shop layouts must essentially include conveniently sited and adequately designed storage facilities for all materials and equipment, including waste materials and scrap. This will allow a smooth flow of work and promote good housekeeping within the workshop, which in itself will improve productivity by the provision of more congenial surroundings.

(b) Improvements to operating conditions

The principles to be applied in analysing operating conditions—or the path of movement of an operator or operators limbs in performing work, are similar to that when considering material flow, i.e. to minimize the distances which have to be travelled between operating points, to eliminate delays, and to avoid any sharp or sudden changes of direction or speed. Movements which will flow easily and smoothly lead to smaller operating times and more efficient execution. Other principles which need to be applied are the characteristics of easy movement and are quoted from No. 22201 of BS 3138 of 1969—

"(i) *Minimum movements.* Movements which, while natural, are the minimum necessary for the job.
(ii) *Simultaneous movements.* Movements in which different limbs are working at the same time.
(iii) *Symmetrical movements.* Movements which are so arranged that they can be performed on the right- and left-hand sides of the body symmetrically about an imaginary plane through the centre of the body.
(iv) *Natural movements.* Movements which make the best use of the shape and arrangement of the part of the body involved.
(v) *Rhythmical movements.* A sequence of movements which induces a natural rhythm when repeated.
(vi) *Habitual movements.* Movements designed, through precise repetition, to become a habit.
(vii) *Continuous movements.* Movements which are smooth and curved and which avoid sharp changes of direction and speed."

When analysing *simo charts*, therefore, it is advantageous to try to combine motions wherever possible, and to attempt to eliminate wasteful therbligs such as "search", "avoidable delay", etc.

When overhauling a work place layout, it is important to locate all the tools, materials and equipment in an area where the operator can reach them, if possible without having to make any abnormal stretching or bending motions so that rhythmical movements can take place.

An improved method cannot be developed entirely on a drawing board or in a laboratory, however. Improvements to operating methods, work place and workshop layout should all be experimented with, if possible on site, to ensure that the theory being developed has a practical solution.

(c) *Improvements to manufacturing methods*

Improvements to processes in manufacture should never be attempted without the assistance of technical specialists. Much preparatory work can be carried out on a logical basis, however, before the specialists are consulted. Considerations of the process would include—

(i) Examination of the sequence of operations with respect to time. Experimenting with combining some of these to reduce movement and handling time and by this, add interest to the job by the introduction of more variety to the work.

(ii) Examination of the methods of performing the work to ensure that there are no unnecessary operations (e.g. an element of work which requires an article to be wrapped in paper at an early stage, subsequently unwrapped for inspection and wrapped yet again later on, the first wrapping being unnecessary for protection purposes).

(iii) Examination of the possibilities of increasing the work content of earlier operations to eliminate later operations.

(iv) Introduction of automatic inspection methods to eliminate a high proportion of specialized inspection routines.

(v) Critical examination of governing elements or "lead" operations (i.e. those which are longer than others in a sequence and tend to govern the work flow). An example of this is given in Chapter 9, Fig 33. By reducing the boiling cycle to half a minute the overall cycle time was reduced by 31·25 per cent.

(vi) The arrangement of machines in such a way that families of similar products can be processed by the same set of machines and therefore follow the same path of movement thereby providing the minimum of delay with the least congestion.

Some technical knowledge may be necessary to allow the machines to be easily set up for different batches of work. One way of doing this is to set some of the machines whilst others are running and then to change over the production line by unplugging and moving out the unwanted machines and moving in and connecting up the pre-set machines, using a small hand-operated hydraulic lift truck for the purpose.

The technique is sometimes known as Group Technology.

(d) Improvement to clerical routines

Generally, improvements to clerical routines need specialized knowledge of the techniques involved. This particularly applies to production control, cost and financial control, maintenance control, etc. Some of the principles applied, however, are constant to all of these such as—

(i) The reduction or elimination of those documents which are distributing information which is of no interest to some of those to whom they are being sent. In this respect, the principle of management by exception should be applied, i.e. that managers can become more effective if the information which is passed to them is restricted to those items which are *not* proceeding according to plan and therefore require management action. To send a wealth of information to a member or members of the management team at different levels to make sure that the right one receives it is both wasteful and ineffective.

(ii) The design of the document system in such a way that information is distributed in time for action to be taken.

(iii) The consideration of adequate filing and data retrieval methods.

(iv) The consideration of using computers for the distribution of information.

(e) Modification to product design

Critical examination of the design of the finished product with a view to reducing material and labour costs without destroying the basic function or sales appeal can often achieve substantial cost reduction.

Some of the considerations involved in this procedure include the following—

(i) Take each part of the product in turn and decide whether it is really necessary for its function, reliability, or aesthetic appeal.

(ii) Similarly examine the product for material content and material specification. Consider alternative materials and construction which will reduce material or labour costs or both, and at the

215

same time either improve the design or leave the essential characteristics unaffected.

(iii) Consider modifications to the product which would facilitate a cheaper tooling method. One example of this would be as follows. By incorporating an extra projecting lug on a casting which is subsequently machined off, location of the part in fixtures is facilitated when earlier operations are carried out, resulting in reduced labour cost and cheaper tools.

(iv) Consider whether a group of products of slightly different design can be replaced by one standard product.

(v) Consider the source from whence the materials are purchased and decide whether the same or similar materials can be obtained more cheaply from other sources of supply.

This process is now known as Value Analysis and is often carried out by a committee who are invited to submit suggestions for improvement however trivial or unconventional. These are then examined and the more interesting ones selected and tried out in practice. The process described under (iv) above is part of a technique known as Variety Reduction.

(f) Reductions to material waste

Material waste can often be reduced by critically examining this aspect at each process carried out, and attempting to produce the same result with less usage of material and consequently less scrap. It is better in this context to strive for more effective use of labour through improvement to the operations and machine settings themselves than through more inspection routines. If scrap work is high, it is usually due to poor manufacturing methods, poor tools or machines, inadequately trained or unsuitable labour, or poor supervision—and not to ineffective inspection. To increase inspection routines will only serve to aggravate the situation without attacking the root cause.

One example of material waste reduction which gave higher quality standards is quoted from a company engaged in producing extruded brass rod.

The brass was first being cast into ingots 6 in. dia. by 6 ft. 6 in. long, and sawn into billets 2 ft long. The 6 in. wastage containing dross and slag was returned to the melting furnace. The billets were next heated to a red heat in a muffle furnace and extruded through hexagon and round dies whilst still hot to produce brass rods of $\frac{1}{2}$ in. to 1 in. section and 60 ft to 250 ft long. At the time of the initial investigation, the billets were being extruded to 1 ft 9 in. length, leaving a 3 in. discard which was subsequently re-melted.

216

The rod was now sawn into 10 ft or 12 ft lengths and any shorter ends re-melted also. The last operation consisted of sizing the section of the 10 ft and 12 ft lengths by drawing them cold through a sizing die.

Re-melt was eventually reduced after a work study investigation by making the following changes—

(i) Sawing three longer billets from the 6 ft 6 in. long ingot, it being found that a 3 in. length was sufficient for dross and slag removal.

(ii) Reducing the discard length after extrusion to $\frac{1}{2}$ in. A specially designed stop was made and fitted to the extrusion press to control this length.

(iii) Preparing sawing tables to ensure that the least amount of extruded rod was sent for re-melt. Operators were trained in the use of these tables.

(iv) Introducing an incentive scheme to encourage saving on re-melt by arranging to pay extra bonus on reduction of scrap combined with higher quality standards. In other words, the operators could earn extra bonus other than that from output alone, if they produced more brass rod of a serviceable standard from the same amount of raw material originally charged to the melting furnace.

One extra inspector was added to the production team to preserve the quality standards.

Development phase

It is essential to carry out practical tests under operating conditions, before implementing any proposed improvements. This may take the form of simulated arrangements using prototype tools and equipment, tests in actual machines, training tests with operators, etc. Because of the differences that will occur between theory and practice, final installation plans should only be drawn up after this phase is complete.

Preparation for installation

Before any attempt is made to install the new methods, supervision and management should be actively consulted and advised. This is of great importance, since the co-operation of foremen is of particular value to the success of any method which will be under their control.

Where there are changes in workshop or workplace layout which involve the installation of special tools and equipment which has first to be manufactured or purchased, a co-ordinated plan should be prepared which takes account of current production demands and the availability of sub-contract or maintenance labour. Project network planning can often be used to advantage in these situations —particularly if the project is of a major nature (see also Chapter 19).

If new documentation is to be introduced, a pilot scheme should first be tested on a limited number of documents produced by office duplicating methods before being finally designed and printed in larger quantities.

Lastly, during introduction, the work study personnel should be in full attendance on site to instruct and help in every possible way until the new methods are running successfully.

Consolidation techniques

Consolidation of new techniques should be achieved by training and instructing operators and supervision, and finally drawing up comprehensive specifications and operating procedures.

In certain instances, it may be necessary to consolidate still further by applying a bonus incentive scheme. This particularly applies where improvements to quality or reductions in material waste are dependent on close attention being given to the work by the operators themselves.

13

Predetermined motion-time systems

As mentioned in the previous chapter, Frank B. Gilbreth proved that physical work can be subdivided into recurring movements which he called "therbligs". For a long while it has been the ambition of many industrial engineers to carry this analysis further and determine standard times for these movements, thus allowing work to be measured without the use of time study.

Such a system must have obvious advantages. For however painstakingly one attempts to explain and demonstrate the soundness of the rating method (see Chapter 2), some authorities remain sceptical. They assert that measured work values obtained by its use are entirely dependent on the judgement of the work study analyst alone and do not accept any statistically based argument. Furthermore, some operators object to the use of a timing device—particularly craftsmen, who are prone to regard submission to its use as somewhat undignified. Another advantage is that building up measured work values from small elements obliges the analyst to examine methods fairly closely and therefore combines method study and work measurement techniques.

There are now many different predetermined motion-time systems in use, but most of these appear to have emanated from either of two basic systems—Work Factor, and Methods-Time Measurement or *MTM*. They were developed from the analysis of a considerable number of ciné films.

The procedure for applying the data is similar for all these systems. The work cycle is observed or mentally visualized, the necessary

219

motions of the hands and other limbs being listed in sequence and classified. Corresponding times for each of the movements are next selected from tables and added together to obtain element times. In most systems, it is then necessary to convert the particular units into basic times by multiplying them by a conversion factor. Standard times can then be determined using exactly the same procedures as outlined in Chapters 4 to 7, i.e. by classifying times and frequencies, adding relaxation allowances to build synthetic data which can finally be used to compute operation times.

The movements of both hands have to be taken into account as well as foot, leg and other body motions, and there are rules governing the relative difficulties of performing certain simultaneous movements, the longer time values being used to compute the work value in these cases.

Work Factor

This was developed by Quick, W. Shear and R. Koehler. It is a proprietary method, and is the registered service mark of the Work Factor system and of the Work Factor Company Incorporated. Much of the original work in developing it was carried out at the Radio Corporation of America between 1934 and 1935.

It is based on the following principles—

(*a*) That body members move at different speeds. The fingers move the fastest, the arms slower, the legs slower still and the trunk the slowest of all.

(*b*) The distance the body members move is proportionate to the time taken, i.e. the longer the distance, the greater the time.

(*c*) Movement becomes more time consuming as difficulties are added to it—for example, when precision of movement is necessary or when extra resistance is present due to added weight, etc.

In what is known as "Detailed Work Factor", times for each movement are obtained by first determining the number of work-factors. These depend on certain conditions, being represented by the following symbols—

W Weight of resistance—because of the weight of the object to be carried, or because of the necessity to expand force in order to overcome resistance.

S Steer (directional control). This is the control necessary due to the performance of a motion in a limited area, or directed towards a small area, and requiring precision.

P Precaution (care). This is the extra care necessary to complete a motion or to maintain a precise control over an operation.

U Change of direction. This is the control required to change direction from a straight line, for example in moving around an obstruction.

D Definite stop. This is the control necessary to stop a motion because the operator wants to stop suddenly, as opposed to the motions being stopped by some physical obstacle such as a bench top or desk.

If one of these elements (or work factors) is present in an operation, it is considered to be one work factor in computing the true allowance. If two are present, this counts as two and so on, up to a maximum of four.

All motions of the body (the arm, the leg, the trunk, the fingers, the hand, the foot and the forearm) are reduced to operations which are described as 400-odd "motion times". These are in unit times of 0·0001 minutes, and are shown on a table for each distance travelled by the body members in performing the work. Where there are no work factors (referred to as "basic") and where there are 1, 2, 3 or 4 work factors, the appropriate values are selected according to the conditions (or work factors) and distances travelled. Similar values are given for various weight factors (i.e. resistance to motion in lb) for both male and female operators, and the degrees of forearm swivel when performing turning operations, together with the torque ratios in lb-ms. The "work-factors" of directional control, care or precaution, change of direction, or definite step would be taken into account in the same way as for distances travelled. There are also additional unit times for walking in free and restricted areas, and for visual focusing.

The tables given in Figs 53 and 54 are detailed work factors suitable for highly repetitive work of very short cycle. There are no allowances for relaxation or delay; these have to be added separately.

Other, modified systems of work factors are—

(*a*) Simplified Work Factor
This is useful for medium quantity work. The time units are in 0·0001 minutes as for detailed work factor.

(*b*) Abbreviated Work Factor
This is intended to be used in jobbing shops, maintenance and construction work, etc. The time units are expressed in 0·005 minutes.

(*c*) Ready Work Factor
This was developed to satisfy the need for a simple method of measuring manual work which can be taught to people who are in functions other than work study (e.g. estimators, etc.). It is expressed in the time-units of 0·001 minutes.

Distance moved (inches)	Basic	Work factors (T) Trunk measured at shoulder				Distance moved (inches)	Basic	Work factors (FH) Finger–Hand measured at finger-tip			
		1	2	3	4			1	2	3	4
1	26	38	49	58	67	1	16	23	29	35	40
2	29	42	53	64	73	2	17	25	32	38	44
3	32	47	60	72	82	3	19	28	36	43	49
4	38	55	70	84	96	4	23	33	42	50	58
5	43	62	79	95	109	Weight (lb) Male		$\frac{2}{3}$	$2\frac{1}{2}$	4 up	—
6	47	68	87	105	120	Weight (lb) Female		$\frac{1}{3}$	$1\frac{1}{4}$	2 up	—
7	51	74	95	114	130	(FT) Foot measured at toe					
8	54	79	101	121	139	1	20	29	37	44	51
9	58	84	107	128	147	2	22	32	40	48	55
10	61	88	113	135	155	3	24	35	45	55	63
11	63	91	118	141	162	4	29	41	53	64	73
12	66	94	123	147	169	Weight (lb) Male		5	22	up —	—
13	68	97	127	153	175	Weight (lb) Female		$2\frac{1}{2}$	11	up —	—
14	71	100	130	158	182	(FS) Forearm swivel measured at knuckle					
15	73	103	133	163	188						
16	75	105	136	167	193	45°	17	22	28	32	3
17	78	108	139	170	199	90°	23	30	37	43	4
18	80	111	142	173	203	135°	28	36	44	52	5
19	82	113	145	176	206	180°	31	40	49	57	6
20	84	116	148	179	209						
Weight (lb) Male	11	58 up				Torque (lb-ins) Male		3	13	up	
Weight (lb) Female	$5\frac{1}{2}$	29 up				Torque (lb-ins) Female		$1\frac{1}{2}$	$6\frac{1}{2}$	up	

Work factor symbols		Walk time				Head turn	
W	Weight or resistance	30 inch paces				45°	40
		Type	1	2	over 2	90°	60
S	Directional control (steer)	General	Analyse from table	260	120 + 80 per pace		
P	Care (precaution)	Restricted		300	120 + 100 per pace	1 Time Unit	
U	Change of direction	Add 100 for 120° to 180° turn at start or finish				= 0·006 second	
		Up steps (8 inches rise, 10 inches flat)			126	= 0·0001 minute	
D	Definite stop	Down steps			100	= 0·00000167 hour	

Fig 53 Detailed Work Factor.* Tables of values for trunk, finger, foot, forearm and head turn movements and walking†

Distance moved (inches)	Work factors				Distance moved (inches)	Work factors			
	1	2	3	4		1	2	3	4
	(A) Arm measured at knuckles					(L) Leg measured at ankle			
	Basic					Basic			
1	18	26	34	40	46	1	21	30	39	46	53
2	20	29	37	44	50	2	23	33	42	51	58
3	22	32	41	50	57	3	26	37	48	57	65
4	26	38	48	58	66	4	30	43	55	66	76
5	29	43	55	65	75	5	34	49	63	75	86
6	32	47	60	72	83	6	37	54	69	83	95
7	35	51	65	78	90	7	40	59	75	90	103
8	38	54	70	84	96	8	43	63	80	96	110
9	40	58	74	89	102	9	46	66	85	102	117
10	42	61	78	93	107	10	48	70	89	107	123
11	44	63	81	98	112	11	50	72	94	112	129
12	46	65	85	102	117	12	52	75	97	117	134
13	47	67	88	105	121	13	54	77	101	121	139
14	49	69	90	109	125	14	56	80	103	125	144
15	51	71	92	113	129	15	58	82	106	130	149
16	52	73	94	115	133	16	60	84	108	133	153
17	54	75	96	118	137	17	62	86	111	135	158
18	55	76	98	120	140	18	63	88	113	137	161
19	56	78	100	122	142	19	65	90	115	140	164
20	58	80	102	124	144	20	67	92	117	142	166
22	61	83	106	128	148	22	70	96	121	147	171
24	63	86	109	131	152	24	73	99	126	151	175
26	66	90	113	135	156	26	75	103	130	155	179
28	68	93	116	139	159	28	78	107	134	159	183
30	70	96	119	142	163	30	81	110	137	163	187
35	76	103	128	151	171	35	87	118	147	173	197
40	81	109	135	159	179	40	93	126	155	182	206
Weight (lb) Male	2	7	13	20	Up	Weight (lb) Male	8	42	Up		
Female	1	3½	6½	10	Up	Female	4	21	Up		

Fig 54 Detailed Work Factor.* Tables of values in time-units for
arm and leg movements†

* Registered trade-mark.
† From *Industrial Engineering Handbook* edited by H. B. Maynard (3rd edn). Copyright © 1951 by McGraw-Hill Inc. Used with permission of McGraw-Hill Book Company.

Predetermined motion-time systems

Methods-time measurement (MTM)

This system was developed in 1940 by Maynard, Stegemerten and Schwab. The data was obtained by film analysis of shop floor operations, mainly on drill-press work at Westinghouse Electric Corporation in the U.S.A. It differentiates amongst motions by describing each, allowing for certain adjustments for different weights carried, obstructions encountered and care necessary in the selection or placement of a body member.

Its development was a direct result of attempting to give time values for "therbligs", although the idea was eventually abandoned, MTM elements being originated in their own right—

MTM Elements	Comparable therbligs
REACH	This is nearest to the "transport empty" therblig.
MOVE	"Transport loaded" and associated therbligs.
GRASP	"Near to grasp" therblig.
POSITION	No comparison.
DISENGAGE	No true equal, but may be compared with "disassemble" in a very loose sense when no recoil is apparent.
RELEASE	Comparable to the therblig "release load".
EYE-TIME	No comparison.
APPLY PRESSURE	There is no directly comparable therblig to this element. It is an application of muscular force to overcome object resistance where little or no movement is expected.

Other elements are included to cover for body, leg and foot motions, including walking.

The element "*REACH*" R is classified into five different cases A, B, C, D and E, with alternatives for the reach elements if the hand is already in motion when these take place. Within these categories, times are listed for distances moved from ¾ inch or less, and 1 inch to 10 inches in inch steps, and from 10 inches to 30 inches in steps of two inches.

The element "*MOVE*" M is similarly classified, but only into three cases A, B and C, with a further alternative for these when the hand is in motion. Different times are also given for distances moved in the same way as for "*REACH*".

224

There is a static (constant) component to allow for taking up a weight with the muscles of the hand and arm prior to moving, and a dynamic component (factor) to compensate for slowing down when an object is heavy.

An element *"TURN"* (T) is also given, with variations for degrees turned modified according to different weight factors present.

"GRASP" (G) is also analysed into different situations according to the nature of the object, the type of grasp, and the conditions under which it is made. Special values are also given for *"REGRASP"* and *"TRANSFER GRASP"*.

"POSITION" P is classified into three types of fit and each of these cases further classified into whether the object is symmetrical, semi-symmetrical or non symmetrical, and easy or difficult to handle.

"RELEASE" RL has two cases, and *"DISENGAGE"* D is classified into "loose", "close" or "tight" fits, and whether the object is easy or difficult to handle.

EYE TRAVEL TIME and *EYE FOCUS ET* and *EF* are also given values as well as *BODY, LEG* and *FOOT MOTIONS*.

The values are expressed in *"TMUs"* (time measurement units) These are 0·00001 hours, 0·0006 minutes or 0·036 seconds. The British Standard rating for these units is $83\frac{1}{3}$. To convert *TMUs* to basic hours, it is therefore necessary to multiply by 5/6 or 0·833. This can be more clearly explained in tabular form as follows—

	TMU	*Hours*	*Minutes*	*Seconds*
MTM unit	1	0·00001	0·0006	0·036
Basic times at 100 British Standard ⎫⎬⎭	—	0·00000833	0·0005	0·030
Factor by which TMUs are to be multiplied to convert to basic times at "100" British Standard	—	0·833 or 5/6	0·0005 or $\frac{1}{2000}$	0·030 or $\frac{3}{100}$

To compute standard operation times it is first necessary to observe or visualize the work and analyse it into its basic *MTM* elements. This forces the observer to carry out a certain amount of method study at the same time, the movements of each hand being considered both separately and in unison, as well as any body, leg and foot movements. The *MTM* Data Card provides analysis which lists those cases of *"REACH"*, *"MOVE"*, *"GRASP"*, *"POSI- TION"* and *"DISENGAGE"* elements which are easily performed by both hands simultaneously, those which can be performed with practice, and those which are difficult to perform—even after long practice. *"TURN"*, *"APPLY PRESSURE"*, *"POSITION"*, *"DIS- ENGAGE"* and *"RELEASE"* are also analysed in this way. The final analysis of the operation is then listed on a chart of the form shown in Fig 61, and the appropriate *TMU* values inserted,

Reach—R

Distance moved (inches)		TMU			Hand in motion		*Case and Description*
	A	B	C or D	E	A	B	
¾ or less	2·0	2·0	2·0	2·0	1·6	1·6	
1	2·5	2·5	3·6	2·4	2·3	2·3	A Reach to object in fixed
2	4·0	4·0	5·9	3·8	3·5	2·7	location or to object in other hand or on which
3	5·3	5·3	7·3	5·3	4·5	3·6	other hand rests
4	6·1	6·4	8·4	6·8	4·9	4·3	
5	6·5	7·8	9·4	7·4	5·3	5·0	B Reach to single object in location which may
6	7·0	8·6	10·1	8·0	5·7	5·7	vary from cycle to
7	7·4	9·3	10·8	8·7	6·1	6·5	cycle
8	7·9	10·1	11·5	9·3	6·5	7·2	
							C Reach to object jumbled
9	8·3	10·8	12·2	9·9	6·9	7·9	with other objects in a
10	8·7	11·5	12·9	10·5	7·3	8·6	group so that search
12	9·6	12·9	14·2	11·8	8·1	10·1	and select occur
14	10·5	14·4	15·6	13·0	8·9	11·5	D Reach to a very small
16	11·4	15·8	17·0	14·2	9·7	12·9	object or where
18	12·3	17·2	18·4	15·5	10·5	14·4	accurate grasp is required
20	13·1	18·6	19·8	16·7	11·3	15·8	
22	14·0	20·1	21·2	18·0	12·1	17·3	E Reach to indefinite
24	14·9	21·5	22·5	19·2	12·9	18·8	location to get hand in position for body balance
26	15·8	22·9	23·9	20·4	13·7	20·2	or next motion or out
28	16·7	24·4	25·3	21·7	14·5	21·7	of way
30	17·5	25·8	26·7	22·9	15·3	23·2	

Fig 55 MTM-1. Tables of values in TMU for element "reach"

these being totalled together with due regard to the effect of simultaneous motions. Conversion to basic times can then be carried out by applying the factors given above, standard times being finally calculated by addition of the appropriate relaxation and contingency allowances.

Times for drilling, metal cutting, cooling, drying or any other such process times must be obtained by time study or from tables of feeds, speeds and other technical tables.

Work standards derived from this method can be within less than ± 1 per cent of those obtained by time study, providing the analyst is properly skilled and experienced in its use.

Second generation systems

Both *MTM* and Work Factor can be used to compute standards of the accuracy which is necessary for repetitive operations. The disadvantage of both these systems, however, is that the computation of standard times can sometimes be a lengthy procedure. Although this can often be justified by the subsequent benefits obtained by the combination of work measurement and method study, it becomes an uneconomical proposition when attempts are made to apply it to non-repetitive work. It is this and the need for the detailed, lengthy and often expensive training necessary before practitioners can become competent in its use which has led to the development of second generation predetermined motion-time systems which, although less accurate, are much simpler in construction and consequently are more easily taught and quicker to apply. Some of these are described below—

(*a*) *Primary Standard Data* (*PSD*)
   This was developed by F. J. Neale of Urwick Orr and Partners. It is a proprietary system.
(*b*) *Master Standard Data* (*MSD*)
   This was developed by Messrs Crossan and Nance of Serge Birn Company. It is also a proprietary system.
(*c*) *Middle-Minute Data* (*MMD*)
   This has been developed by P.A. Management Consultants and is also a proprietary system.
(*d*) *Simplified Predetermined Motion Time Data* (*SPMTS*)
   This was developed by Russel Currie at ICI. It is not a proprietary system. Mr Currie wrote a book describing this system (q.v.) which is extremely detailed in its explanations. It describes a kit which can be constructed to enable the reader to teach himself the method.
(*e*) *General Purpose Data* (*GPD*)
   Developed by the American MTM Association in 1962.

227

Predetermined motion-time systems

Move—M

Distance moved (inches)	TMU A	B	C	Hand in motion B	Case and description	Wt (lb) up to	Dynamic factor	Static constant TMU
¾ or less	2·0	2·0	2·0	1·7		2·5	1·00	0
1	2·5	2·9	3·4	2·3				
2	3·6	4·6	5·2	2·9	A Move object to other hand or against stop	7·5	1·06	2·2
3	4·9	5·7	6·7	3·6				
4	6·1	6·9	8·0	4·3				
5	7·3	8·0	9·2	5·0		12·5	1·11	3·9
6	8·1	8·9	10·3	5·7				
7	8·9	9·7	11·1	6·5		17·5	1·17	5·6
8	9·7	10·6	11·8	7·2				
9	10·5	11·5	12·7	7·9	B Move object to approximate or indefinite location	22·5	1·22	7·4
10	11·3	12·2	13·5	8·6				
12	12·9	13·4	15·2	10·0		27·5	1·28	9·1
14	14·4	14·6	16·9	11·4				
16	16·0	15·8	18·7	12·8		32·5	1·33	10·8
18	17·6	17·0	20·4	14·2				
20	19·2	18·2	22·1	15·6	C Move object to exact location	37·5	1·39	12·5
22	20·8	19·4	23·8	17·0				
24	22·4	20·6	25·5	18·4		42·5	1·44	14·3
26	24·0	21·8	27·3	19·8				
28	25·5	23·1	29·0	21·2				
30	27·1	24·3	30·7	22·7		47·5	1·50	16·0

Fig 56 MTM-1. Tables of values in TMU for element "move"

Turn—T

Weight	TMU for degrees turned 30°	45°	60°	75°	90°	105°	120°	135°	150°	165°	180°
Small 0 to 2 lb	2·8	3·5	4·1	4·8	5·4	6·1	6·8	7·4	8·1	8·7	9·4
Medium 2·1 to 10 lb	4·4	5·5	6·5	7·5	8·5	9·6	10·6	11·6	12·7	13·7	14·8
Large 10·1 to 35 lb	8·4	10·5	12·3	14·4	16·2	18·3	20·4	22·2	24·3	26·1	28·2

Fig 57A MTM-1. Table of values in TMU for element "turn"

Apply Pressure—AP

Apply pressure case 1—16·2 TMU. Apply pressure case 2—10·6 TMU	
Apply force (AF) = 1·0 + (0·3 × lb) TMU for up to 10 lb = 4·0 TMU max for 10 lb and over	

Dwell minimum (DM) = 4·2 TMU	*Release force* (RLF) = 3·0 TMU
AP = AF + DM + RLF	APB = AP + G2

Grasp—G

Case	TMU	*Description*
1A	2·0	Small, medium or large object by itself, easily grasped
1B	3·5	Very small object, or object lying close against a flat surface
1C1	7·3	Interference with grasp on bottom and one side of nearly cylindrical object. Diameter larger than $\frac{1}{2}''$ but not over 1 $''$
1C2	8·7	Interference with grasp on bottom and one side of nearly cylindrical object. Diameter $\frac{1}{4}''$ to $\frac{1}{2}''$
1C3	10·8	Interference with grasp on bottom and one side of nearly cylindrical object. Diameter less than $\frac{1}{4}''$
2	5·6	Regrasp
3	5·6	Transfer grasp
4A	7·3	Object jumbled with other objects so search and select occur. Larger than $1'' \times 1'' \times 1''$
4B	9·1	Object jumbled with other objects so search and select occur. $\frac{1}{4}'' \times \frac{1}{4}'' \times \frac{1}{8}''$ to $1'' \times 1'' \times 1''$
4C	12·9	Object jumbled with other objects so search and select occur. Smaller than $\frac{1}{4}'' \times \frac{1}{4}'' \times \frac{1}{8}''$
5	0	Contact grasp

Fig 57B MTM-1. Tables of values in TMU for elements "apply pressure" and "grasp"

Position—P

	Class of fit	Symmetry	Easy to handle	Difficult to handle
1 Loose	No pressure required	S SS NS	5·6 9·1 10·4	11·2 14·7 16·0
2 Close	Light pressure required	S SS NS	16·2 19·7 21·0	21·8 25·3 26·6
3 Exact	Heavy pressure required	S SS NS	43·0 46·5 47·8	48·6 52·1 53·4

Distance moved to engage 1″ or less

Release—RL

Case	TMU	Description
1	2·0	Normal release performed by opening fingers as independent motion
2	0	Contact release

Disengage—D

Class of fit	Easy to handle	Difficult to handle
1 *Loose*—Very slight effort blends with subsequent move	4·0	5·7
2 *Close*—Normal effort slight recoil	7·5	11·8
3 *Tight*—Considerable effort, hand recoils markedly	22·9	34·7

Eye travel time and eye focus—ET and EF

Eye travel time = $15·2 \times T/D$ TMU with a maximum value of 20 TMU

where T = the distance between points from and to which the eye travels
D = the perpendicular distance from the eye to the line of travel T

Eye focus time = 7·3 TMU

Fig 58 MTM-1. Tables of values in TMU for elements "position", "release" "disengage", "eye travel" and "eye focus"

Body, leg and foot motions

Description	Symbol	Distance	TMU
Foot motion			
Hinged at ankle	FM	Up to 4″	8·5
With heavy pressure	FMP		19·1
Leg or foreleg motion	LM—	Up to 6″	7·1
		Each add ½″	1·2
Sidestep			
Case 1—Complete when	SS—C1	Less than 12″	Use reach or
leading leg			move time
contacts floor		12″	17·0
		Each add ½″	0·6
Case 2—Lagging leg must	SS—C2	12″	34·1
contact floor		Each add ½″	1·1
before next			
motion can be			
made			
Bend, stoop, or kneel on	B, S, KOK		29·0
one knee			
Arise	AB, AS, AKOK		31·9
Kneel on floor—both knees	KBK		69·4
Arise	AKBK		76·7
Sit	SIT		34·7
Stand from sitting position	STD		43·3
Turn body 45 to 90 degrees			
Case 1—Complete when	TBC1		18·6
leading leg			
contacts floor			
Case 2—Lagging leg must	TBC2		37·2
contact floor			
before next			
motion can be			
made			
Walk	W-FT	Per foot	5·3
Walk	W-P	Per pace	15·0
Walk obstructed	W-PO	Per pace	17·0

Fig 59A MTM-1. Table of values in TMU for "body, leg and foot motions"

Crank—light resistance—C*

Diameter of cranking—inches	TMU per intermediate revolution (T)
1	8·5
2	9·7
3	10·6
4	11·4
5	12·1
6	12·7
7	13·2
8	13·6
9	14·0
10	14·4
12	15·0
14	15·5
16	16·0
18	16·4
20	16·7

For resistances other than light, use the weight allowance table in MOVE

Fig 59B MTM-1. Table of values in TMU for "crank"

* These values are still under research.

(f) *Modular Arrangement of Predetermined Time Standard*
   (*MODAPTS*)
   This is a proprietary system developed by G. C. Heyde at
   Unilever in Australia.
(g) *Universal Standard Data* (*USD*)
   Another proprietary system developed by H. B. Maynard and Co.
(h) *Integrated Standard Data* (*ISD*)
   This was developed by IBM.

All these systems are highly effective in use although they are
largely based on different principles. But some of them are proprietary
systems, and the precise methods used in the development of many
of them are not generally made known. These are considered to be
drawbacks by some authorities who believe that a universally acceptable second-generation system should have openly available back-up
data which is not only capable of substantiating it, but also of developing it in the future.

A system which does overcome all of these disadvantages is
*MTM-*2. This was developed by the "Standing Committee" for

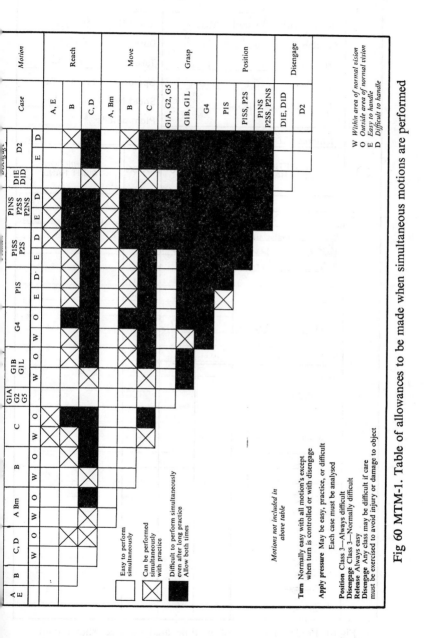

Fig 60 MTM-1. Table of allowances to be made when simultaneous motions are performed

P.M.T.S. ANALYSIS SHEET

Department	Work study	Study No.	2075
Section	Experimental	Date	2.8.71
Product	Fountain pen	Taken by	S.T.C.
Operation	Place cap on	Operator	B. Hall
	fountain pen	Sheet No.	1 of 1
		Plant details	

Remarks   Example only

DESCRIPTION— LEFT HAND	MOTION	TMU	MOTION	DESCRIPTION— RIGHT HAND
		14	GB30	Get pen from desk
		11	PA30	Put pen to assembly position
Get cap from desk	GC30	23	(R)	Re-grasp pen
Put cap on pen	PC45	36		
Aside hand to rest	PA30	11		
		11	PA30	
		11	PA30	
TOTAL T.M.U's.		117		
= 5·85 basic centiminutes				

Fig 61 Predetermined motion time analysis sheet shown completed with an MTM-2 pattern for placing a cap on a fountain pen. The "regrasp" element is limited out

applied research of the International MTM Directorate and is based on *MTM* (which is now known as *MTM*-1 to distinguish it from this second generation system).

MTM-2

The data for this system was obtained by collecting a considerable amount of details of work operations from different industries* and subjecting this to analysis by computer, basing the analysis on *MTM*-1 principles.

By combining, averaging, substituting and eliminating various of the *MTM*-1 motions, *MTM*-2 finally emerged in 1965.

Primarily, the development was possible because certain of the *MTM*-1 motions were often performed together. The list was reduced to 9 motions only—

MTM-2 motion	*Derivation from MTM-1 motions*
1 *GET G*	Generally derived from combining weighted proportions of **REACH**, **GRASP** and **MOVE**.
2 *PUT P*	Developed from combining weighted averages of MOVE and POSITION.
3 *APPLY PRESSURE A*	Generally developed by weighted averaging APPLY PRESSURE (A and B).
4 *RE-GRASP R*	Developed from REGRASP.
5 *EYE ACTION E*	Developed from averaging the EYE TRAVEL TIME and EYE FOCUS.
6 *CRANK C*	A newly developed motion based on weighted averages.
7 *STEP S* 8 *FOOT MOTION F*	Based on averaging similar elements in MTM-1 (W, SS, TB, FM, LM).
9 *BEND and ARISE B*	Developed by combination and average of the BEND, STOOP, KNEEL and ARISE elements

The *MTM*-2 motion "*GET*" is analysed in 3 types or cases, **A**, **B** and **C**—

GA = No grasping motion.
GB = Single grasping motion.
GC = More than one grasping motion.

Each of these types of grasp is further categorized into distance codes of 5, 15, 30, 45, and over 45 cm length (symbolized by the figure 80).

* In the UK, USA and SWEDEN.

The motion "*PUT*" is also analysed in 3 cases—

PA = Continuously smooth motion.
PB = When there is not a smooth motion but no obvious correcting motions.
PC = When there are obvious correcting motions.

In each category of "*GET*" and "*PUT*", there are added factors for weight or resistance to motion (*GW* and *PW* respectively).

There is only one value given in the system for each of the motions "apply pressure", "re-grasp", "eye-action", "crank", "step", "foot motion" and "bend and arise".

Limit of reach in centimetres	Get			Put		
Code	GA	GB	GC	PA	PB	PC
−5	3	7	14	3	10	21
−15	6	10	19	6	15	26
−30	9	14	23	11	19	30
−45	13	18	27	15	24	36
−80	17	23	32	20	30	41

GE (Get weight) GWI—1Kg   PW (Put weight) PWI—5Kg.

A (*Apply pressure*)	R (*Regrasp*)	E (*Eye action*)	C (*Crank*)	S (*Step*)	F (*Foot motion*)	B (*Bend & arise*)
14	6	7	15	18	9	61

Fig 62 MTM-2. Tables of all values used in TMU (there are 32 in all). MTM-2 can be combined with MTM-1 if desired as the units and derivations of data are the same

In all, there are only 39 values. This makes the system extremely simple to teach and to use. The whole principle and method can be explained in a few days, and standards can be computed four times as fast as when using *MTM*-1. The data from *MTM*-1 and *MTM*-2 can furthermore be interchanged as the units are the same (*TMUs*), and the motion pattern used will be similar as they are based on the same principles.

MTM-3

This is a third generation *MTM* system which was announced in 1970. It has been developed by further simplifying the basic *MTM*-1 motions and their variables. The values given in the tables are in *TMUs*—

Limit of reach in centimetres	*Handle*		*Transport*	
Code	HA	HB	TA	TB
−15	18	34	7	21
−80	34	48	16	29

SF (*Steps*)	B (*Bend and arise*)
18	61

The motions used are based on theory that often, many *MTM*-2 motions are performed together—

The hand motions Handle (H) and Transport (T) are classified into cases A and B according to the degree of control necessary to place the object, and into motion–distance classifications of 15 and 80 cm.

237

MTM-3 is intended for use where the speed of analysis is more important than actual precision in the determination of time values. It is most suitable, therefore, where a product is made in small quantities, with the methods used and motion distances encountered tending to vary considerably, thus giving variation in work content for the same basic jobs. It is therefore very useful for measuring certain types of jobbing, tool room and maintenance work.

Specialized data systems

The method of using predetermined motion time systems is as follows—

(*a*) Observe the operation and record the method, detailing the activity of each hand, and other limbs—where these are used in doing the work.

(*b*) Enter the time standards from the basic data tables forming *work elements*. These work elements will be those as described in Chapter 3.

If the procedures as explained in Chapters 4 to 7 are then followed, there will be produced a series of elements and elemental operations —some of which will vary according to the characteristics of the product, process, material, etc. and others which will remain constant.

Some elements will recur regularly in similar fields in different industries. For example, a typist in a heavy engineering works will be performing common elements to her counterpart in a publisher's office. Using a predetermined motion time system, these would be revealed as a constant combinations of elemental movements. Such combinations are now termed "*DATA BLOCKS*". If one takes the principle further (as explained in Chapters 6 and 7), to a lesser degree common combinations of data blocks known as "motion sequences" will be seen to exist. For example, in offices everywhere the element "procure sheet of typing paper" must occur regularly and can be formed into a "data block" by synthesis from basic *PMT* data.

Taking this a stage further, the motion sequence "procure sheet of typing paper, place in typewriter and adjust for position ready for typing" must also be frequently encountered. Building up in this way can gradually form specialist predetermined motion time system data which are peculiar to certain fields of activity. Much work has already been carried out in this field using *MTM* as the basis and some examples are given below—

(*a*) Master clerical data and universal office controls. These are for use in general office work.

(*b*) Universal maintenance standards—used for maintenance work.

(*c*) Standard sewing data, used in the clothing industry.

It must be pointed out, however, that it is equally possible and feasible (indeed it has been done many times in the past) to develop such and similar data by time study, using the principles outlined in Chapters 3 to 11.

Tape data analysis

One of the encumbrances when using *PMT* systems—particularly on non-repetitive work when there is no "second chance" to note elements which have been missed, is the difficulty of writing down intelligibly what is happening with sufficient speed. This can be overcome by using a small, portable tape recorder into which the analyst speaks as the action takes place. The sequences can be played back in the office, the analyst stopping the recorder periodically to allow them to be written down legibly.

For this purpose, a shortened descriptive phraseology must be used, so that each particular motion can be recorded the instant it occurs.

14

Incentives

Introduction

There are many factors which influence the will to work. Some of them can be listed as follows—

(a) Interest in the work.

(b) A sense of achievement. This includes pride in the product and the organization.

(c) Congenial surroundings. Where this is indoors it includes a clean, adequately spaced, well-ventilated, well-lit and properly heated (and cooled) workshop (or office), as well as a harmonious atmosphere of working.

(d) A sense of belonging to a team or group.

(e) A feeling that the welfare of the individual matters.

(f) The basic need for survival.

Interest in the work is undoubtedly high on the list for a great deal of apathy can develop from boredom. One of the causes of boredom is highly repetitive work that gives little or no sense of achievement—except in the sheer numbers that can be produced. Attempts to revive interest in the job is now being tried out by a technique known as job enlargement. This is a reverse of the conventional approach. Instead of breaking a job down into small operations which are required to be performed repetitively, the work is arranged so that each operator produces a complete article—say assembles a propelling pencil—instead of performing only a small part of the job. This is meeting with some success, since it includes pride in the

240

work, as well as eliminating some of the problems of line balance which occur on team-work (see Chapter 9).

Congenial surroundings, a sense of belonging, and welfare are also very important factors. They in themselves tend to give better industrial relations which are so important to high morale and high productivity.

But these cannot succeed in themselves. In the final analysis, the strongest incentive of all is the basic need for survival. Financial incentive asserts that need, and when properly designed and administered, will undoubtedly induce operators to work at very high performances.

One of the greatest sources of discontent is a sense of injustice—and operators usually have a strong sense of justice. They will always maintain that two people working in the same job and producing at the same rate should receive equal pay. Which is really saying that if one produces more than another in the same job he should receive more pay. This is the basis of output incentives—to see that justice is done by rewarding the hardest worker with higher pay in his own grade. For to deny workers the right to earn more if they increase their effort is in a sense unjust; those seeking higher standards of living will certainly see it that way.

It is not with the basic idea of financial incentives that workers show discontent but with the unfairnesses that creep into those poorly administered reward schemes and with the resulting annoying fluctuations in earnings which do not reflect the same fluctuation of effort.

To succeed at all, therefore, financial incentives must be logically based, fairly conceived and carefully administered. Most important of all—they will not succeed in their purpose if the other influences of welfare, good administration, fair treatment and proper working conditions are not given equal attention. The best and most lasting results are obtained when all these factors are taken continuously into account. This is not only common sense—it is good management.

Incentives for direct labour

There are many different systems of payment by results for direct labour, but the most important types are as follows—

(*a*) Directly proportionate output incentives.
(*b*) Differential output incentives.
(*c*) Stabilized output incentives.
(*d*) Special incentives.

These are now described in more detail.

(a) *Directly proportionate output incentives*

These reward operators in direct proportion to output. They are usually in either of two forms—

(i) Piecework

This is probably the oldest form of incentive payment systems. Each unit or piece is given a fixed price and earnings depend on the quantities produced. Nowadays, piecework systems guarantee a minimum rate per hour, wages rising above this level in direct proportion to output. Compensating payments are also made for waiting time and for work which has not been priced. The system is shown graphically in Fig 63.

The main advantage of this system is its simplicity, and operators can easily understand it. It is difficult to maintain it in its original form if there is a wage award, however, as there may be considerable clerical work involved in amending the piecework prices.

An example of wage calculations using a piecework incentive is as follows—

Output	10,000 of job A at £0·1 per 100 pieces
	18,000 of job B at £0·05 per 100 pieces
Waiting time	2 hours
Time working on jobs not priced	1 hour
Total time in attendance	40 hours
Payrate for waiting time, jobs not priced and guaranteed minimum	£0·4 per hour

Fig 63 Directly proportionate and differential output incentives. Base wage and total earnings levels can be set differently to suit local conditions

A. Piecework. The number of pieces produced determines the total earnings above a guaranteed minimum base wage

B. Standard time system. Total earnings above a guaranteed minimum base wage vary directly with performance. Target earnings are at 100 BSI

C. Modified or "geared" standard time systems where earnings are not in true proportion to effort but target earnings remain at 100 BSI

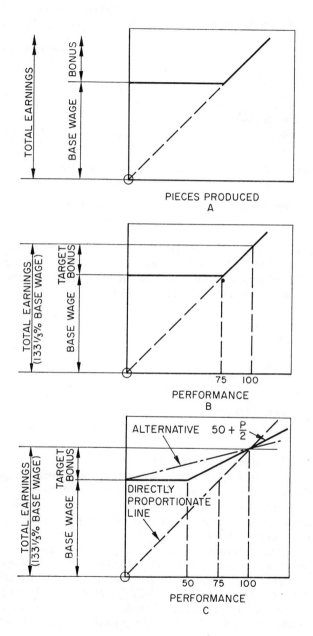

PIECES PRODUCED
A

PERFORMANCE
B

PERFORMANCE
C

243

<table>
<tr><td>*Piecework earnings*</td><td>$P = \dfrac{(10,000 \times £0{\cdot}1)}{100} +$</td></tr>
</table>

Piecework earnings

$$P = \frac{(10,000 \times £0{\cdot}1)}{100} +$$

$$\frac{(18,000 \times £0{\cdot}05)}{100}$$

$$= £19$$

Payment for waiting time and jobs not priced

$$A = (1 + 2) \times £0{\cdot}4$$
$$= £1{\cdot}2$$

Gross earnings

$$P + A = £19 + £1{\cdot}2$$
$$= £20{\cdot}2$$

Minimum wage

$$w = 40 \times £0{\cdot}4$$
$$= £16$$

If the gross earnings are less than the minimum wage they must be made up to it, the difference being known as "make-up".

Sometimes, calculations show the guaranteed minimum plus a bonus. In the example this would be—

Minimum or base wage £16

Bonus (gross earnings less the minimum wage) £4·2

Gross earnings (as before) £20·2

(ii) Standard time system (Fig 63)
This is based on the same principle as piecework except that the values are issued in units of time, i.e. quantities per hour, or standard or allowed time per number of pieces. There is usually a guaranteed minimum rate and rates of pay for waiting time and work not rated.

Operators usually prefer this type of incentive to straight piecework as they can better judge their earnings levels, particularly if they **know** those quantities per hour which will give them a target rate of pay. Adjustment of wage levels in line with agreed wage increases is simpler, as it is merely confined to altering measured work, waiting time, unmeasured work and minimum wage rates.

An example of wages calculations for an incentive system using standard times is given below—

Output	1,000 of job A at 11 standard hours per 1,000 pieces
	2,000 of job B at 10 standard hours per 1,000 pieces
Waiting time	2½ hours
Time working on work not measured	1½ hours
Total time in attendance	40 hours
Pay rate for waiting time, jobs not priced and guaranteed minimum	£0·45 per hour
Pay rate per standard hour produced	£0·6

Standard hours produced

$$H = \frac{(1,000 \times 11)}{1,000} +$$

$$\frac{(2,000 \times 10)}{1,000}$$

$$= 31$$

Payment for standard hours produced

$$P = 31 \times £0·6$$
$$= £18·6$$

Payment for waiting time and jobs not priced

$$A = (2\tfrac{1}{2} + 1\tfrac{1}{2}) \times £0·45$$
$$= £1·8$$

Gross earnings

$$P + A = £18·6 + £1·8$$
$$= £20·4$$

An alternative method of arriving at earnings is to calculate the operator performance and obtain the rate per hour from a table constructed for this purpose. In the example this would be carried out as follows—

$$\text{Operator performance} = \frac{100 \times \text{standard hours produced}}{\text{Actual hours taken to produce them}}$$

$$= \frac{100 \times 31}{40 - (1\tfrac{1}{2} + 2\tfrac{1}{2})}$$

$$= \frac{3100}{36}$$

$$= 86 \text{ BSI}$$

The bonus rate per hour for this (from table Fig 64) $= £0\cdot066$ per hour

Bonus for standard hours produced $= 36 \times £0\cdot066$
$= £2\cdot4$

Base wage $= 40 \times £0\cdot45$
$= £18$

Gross earnings $=$ Base Wage $+$ Bonus
$= £18 + £2\cdot4$
$= £20\cdot4$ as before

The table (Fig 64) is constructed from the formula—

Bonus per hour $= \dfrac{\text{operator performance} \times \text{Base wage per hour}}{75}$
$-$ (Base wage per hour)

(b) Differential output incentives

These reward operators according to output, but are not in the same proportion for all levels. They can take a variety of forms, examples being as follows—

(i) Differential piecework

In this type of incentive, each unit or piece is given a price which varies with the rate at which they are produced. For example, piecework prices for the same job could be defined as follows—

Up to 200 per hour	£0·20 per 100 pieces
201–220 per hour	£0·22 per 100 pieces
221–250 per hour	£0·25 per 100 pieces

75 or less	Nil	90	£0·090	105	£0·180
76	£0·006	91	£0·096	106	£0·186
77	£0·012	92	£0·102	107	£0·192
78	£0·018	93	£0·108	108	£0·198
79	£0·024	94	£0·114	109	£0·204
80	£0·030	95	£0·120	110	£0·210
81	£0·036	96	£0·126	111	£0·216
82	£0·042	97	£0·132	112	£0·222
83	£0·048	98	£0·138	113	£0·228
84	£0·054	99	£0·144	114	£0·234
85	£0·060	100	£0·150	115	£0·240
86	£0·066	101	£0·156	116	£0·246
87	£0·072	102	£0·162	117	£0·252
88	£0·078	103	£0·168	118	£0·258
89	£0·084	104	£0·174	119	£0·264

Fig 64 Specimen table showing bonuses per hour against various performances

The system is used to encourage operators to work at the highest possible rate.

It has many disadvantages. It is complicated to administer, difficult for operators to understand and generally causes a lowering of quality standards. Because of its unpopularity it is becoming increasingly uncommon.

(ii) Modified standard time systems

In certain processes, the work content of jobs may vary due to unavoidable conditions. Some materials become more difficult to process in humid conditions. Sand takes a greater effort to shovel when it is wet rather than dry. Under a normal standard

247

time system earnings could fluctuate considerably and cause discontent. One of the ways of stabilizing these it to calculate them on the basis of an incentive as shown graphically in Fig 63. This is sometimes referred to as 50 + P/2 incentive, as the equivalent performance can be calculated from this formula. For example, if an operator achieved a performance of 80 by producing 8 standard hours in 10 actual hours, payment received would be equivalent to—

$$50 + \frac{80}{2}$$

$$= 90 \text{ performance}$$

if the operator had been on the normal standard time incentive. In other words, if the gross rate for 100 performance was £0·5 per hour, the operator would receive £0·45 per hour for this performance.

If, however, the operator produced 12 standard hours in 10 actual hours (i.e. a 120 performance) payment would be equivalent to—

$$50 + \frac{120}{2}$$

$$= 110 \text{ performance}$$

(£0·55 per hour in the example)

From this it can be seen that if the performance is below 100, the earnings paid will be made up to a level which is half-way between actual performance and standard, whereas if the performance is higher than standard, earnings are reduced by half the difference between actual performance and standard.

The incentive has the disadvantage of encouraging operators to maintain their performance to around 90 (to receive equivalent earnings to 95 per cent of standard) as they do not consider it worth while to make the extra effort to attain target.

A variation of this type of incentive is also shown in Fig 63.

(iii) Limiting differential systems

These systems are designed to limit earnings to a ceiling figure and are used when work cannot be measured accurately and has to be estimated. The best known system of this type is known as the Rowan system which is based on the formula—

$$\text{Bonus payment} = \text{Base wage} \times \frac{\text{Time allowed—time taken}}{\text{Time allowed}}$$

Values are issued as "allowed times". It is shown graphically in Fig 65A. Bonus earnings can never be equal to the base wage. A useful modification of this system is shown in Fig 65B. A chart is drawn with percentage of basic wages paid as bonus plotted against the values of $\dfrac{\text{Time taken}}{\text{Standard time}}$

A straight line is then drawn to join two values which are chosen as maxima.

In the example shown in Fig 65B, maximum bonus would not exceed 73 per cent of basic wage, but bonus would be earned by the operator as long as the time taken did not exceed $1\frac{1}{2}$ times the standard time; and when the time taken is equal to the standard time, target bonus would be paid at the rate of 25 per cent of basic wage.

Maximum values which determine bonus payments can be chosen to suit circumstances.

The advantage of this type of incentive is that earnings limits can be set beforehand and never get out of hand. Its disadvantage is that it destroys the incentive if the standard or allowed times become loose causing discontent amongst operators.

Earnings calculations are best made by calculating performances and obtaining the equivalent rates per hour from a table as explained under (*a*) (ii) (see Fig 64).

Because of their relative ineffectiveness and unpopularity, and because of the difficulty in standardizing labour cost, differential output incentives are not to be recommended for general use.

(*c*) *Stabilized output incentives*
One of the aggravating influences of financial incentives is that they tend to cause wide fluctuations in earnings, particularly if they are poorly designed and administered. This in its turn can be the source of discontent amongst labour, a great deal of time being lost settling disputes and generally administering the scheme. Stabilized incentives can to some extent overcome these disadvantages and, although some of the impetus of the incentive may be lost, the overall effect can be beneficial in many instances. Generally, there are three ways of achieving this—

(i) Measured Daywork
This system consists of specifying a target level of output and other conditions in order to qualify for a fixed earnings level per hour. For example, an operator purely on timework (i.e. fixed rate per hour irrespective of output) will be paid a higher

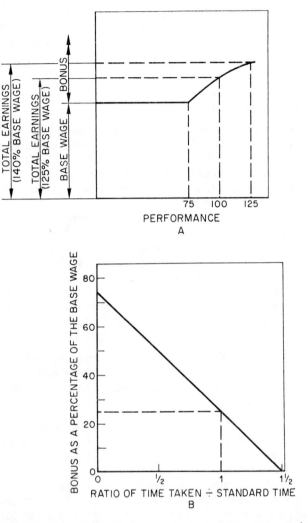

Fig 65 Limiting differential bonus systems. Base wage, earning levels and shapes of curve can be modified to suit the prevailing conditions

A. Bonus system which limits earnings by decreasing the rate of payments as the performance rises

B. Bonus system similar to that given above but expressed in a different form. Limits can be chosen to suit conditions

rate if he agrees to produce at a certain level of performance. In order to qualify, he may be required to maintain this level for a period of (say) one month. Thereafter, his performance is calculated regularly and he can retain his grade and therefore his earnings level as long as he maintains his performance level.

For example, it may be decided that in order to qualify for a pay rate of £0·6 per hour, operators must maintain a true performance of not less than 90 BSI for a period of one month. From then on regular checks are made on performances each week. Should any operator fall below this target over one month, he is given a warning. If he continues to default over the next consecutive month then he is downgraded to base rate (which may be £0·45 per hour) and paid as an ordinary time worker

Calculation of wages is simple in this system and it is generally popular as it eliminates fluctuations in pay. It does, however, demand a generally higher standard of supervision because an even flow of work is essential to ensure that the performance standards can be maintained.

(ii) Premium Pay Plan.

This is a refinement of the measure daywork system and introduces an incentive whilst still preserving a constant earnings level. It also acknowledges that optimum performances of individual operators will vary, and allows them to be graded according to their capabilities. It furthermore prevents earnings from rising disproportionately over a period of time, particularly if the incentive conditions are periodically subjected to audit.

The scheme is shown graphically in Fig 66.

As an *example* of its use, five output levels or grades could be set as follows—

Grade	Performance levels	Rates per hour
1	over 75	£0·48
2	over 85	£0·54
3	over 95	£0·60
4	over 105	£0·66
5	over 115	£0·72

Operators are given the option of deciding which grade they would choose to work in, the qualifying period being set at one month (or more) of continuous maintenance of the performance level. The operator is then paid the appropriate rate per hour

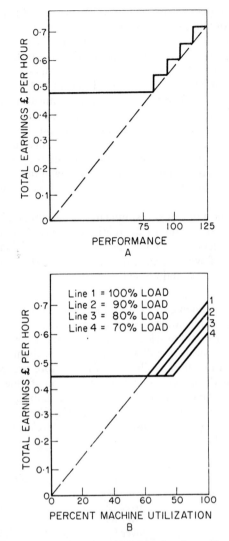

Fig 66 Graphs showing methods of paying incentives based on graded performance and machine or plant utilization

A. Graded performance scheme or "premium pay plan". Earnings are stabilized in payment "bands". The graph can be modified to suit conditions

B. Type of incentive for machine or plant utilization for varying loads allocated

for all the hours he works if the output level is consistently maintained. Should it fall below for one month, however, the operator is given a warning, and if it is still not maintained after a second month, he is automatically downgraded to the next lower level. Any operator can apply for upgrading at any time and therefore the incentive is preserved, but the rate of pay remains static whilst the operator remains in the selected grade.

The system is a little complex to operate although relatively simple if the payroll is produced by a computer which is programmed to "downgrade" and "upgrade" according to the specified conditions. Apart from stabilizing the earnings, it has the advantage of being able to accommodate unavoidable fluctuations in work content without causing similar fluctuations in pay rates.

(iii) Group incentives

Where operators are performing a specific task as a member of a team or gang it is often impracticable to pay incentive bonus individually. In these cases bonus can be calculated as a group and divided amongst the individual members according to the hours that they work.

Any of the types of incentive so far described can be used for this purpose. This method has the effect of smoothing out unavoidable fluctuations in work content which is useful to provide stability of earnings, but the incentive effect may be weakened if some members of a team tend to rely on the remainder to keep up the output and thus cause discontent.

For this reason, group incentives should only be used when individual schemes are difficult to apply, such as on flowline teams or maintenance gangs. However, if the leader is a foreman or chargehand and has a strong personality, performance throughout the team may be kept at a high level.

(*d*) *Special incentives*

Incentives need not necessarily direct operator effort towards improvement in output. Often, it is preferable to maintain high quality standards or reduce material waste. Special needs may also arise in process-controlled work and inspection services. In these cases, incentives should be carefully designed to achieve the desired result. Examples of some of these are given below—

(i) Incentives for machine utilization

Where high machine utilization has a greater effect on cost than high utilization of labour, earnings should be based on the performance of the machine or process. Where it can be arranged that the number of machines or processes allocated represents standard performance when the operator maintains them at optimum machine utilization, payment could simply be arranged to vary in direct proportion to the performance of the machine. But sometimes, operators may be required to control a number of machines which does not represent a full load in this way. A type of incentive which caters for such conditions is shown graphically in Fig 66B.

For example, if an operator could be allocated, say, 10 machines which represented a performance of 100 BSI, to maintain these at 90 per cent utilization; then when a full load of 10 machines was being operated (i.e. the load is 100 per cent) earnings would be related to line 1 on the graph Fig 66B. For lesser loads of 9 machines (90 per cent load); 8 machines, 7 machines, etc., payment would be related to lines 2, 3 and 4 respectively. The following table illustrates the principle of this and is taken from Fig 66 (in practice, smaller intervals of performance and machine utilization—in steps of one—would be shown)—

Per cent machine utilization	Allocated load			
	100%	90%	80%	70%
100	£0·700	£0·670	£0·630	£0·600
90	£0·630	£0·603	£0·567	£0·540
80	£0·560	£0·536	£0·504	£0·480
70	£0·490	£0·469	£0·450	£0·450
60	£0·450	£0·450	£0·450	£0·450
50	£0·450	£0·450	£0·450	£0·450

Table showing gross earnings levels per hour for different loads and machine utilizations

Thus, the earnings levels are less for the same per cent machine utilization if the load is lessened, but not in the same proportion as the available work content.

(ii) Incentives to improve quality standards

One of the criticisms often levelled at output incentives is that they tend to reduce quality standards, and in some cases the benefits resulting from increased output are more than offset by the loss due to increased scrap. This is mainly an unfair criticism, since one of the main conditions of bonus payment under an output incentive should be that credit will only be given for work of acceptable quality. Nevertheless, very often inspection services have to be reinforced to protect quality standards when they are installed. Also, in those industries where quality is of paramount importance (manufacturers of fine pottery, *objets d'art*, etc.), conventional output incentives could certainly do a great deal of damage.

In these circumstances, therefore, specially designed incentives which will influence operators to maintain high quality standards may be necessary.

One way of doing this is to pay part of the bonus on quantities produced using one of the methods so far described, and part on condition that no faulty work is produced, with severe reduction of bonus for each faulty piece found. Proportions of bonus paid would have to be arranged according to the relative importance to high quality and quantity.

(iii) Incentives to reduce material waste

Greater care in the handling and usage of material may result in substantial cost reductions, particularly in those industries where rare metals are used, or where costly processes are being employed. Examples occur in chromium plating, where conservation in the use of nickel can reduce operating costs to a high degree, and in metal casting, forging and extrusion processes, where reduction in wastage can mean less costs spent on remelting.

One method of devising such an incentive is to arrange to pay part of the bonus on quantities produced with an additional, larger part, which can be earned by maintaining the scrap (re-melt, etc.) at a low level, reducing this part of the bonus proportionate to the increase in scrap.

A specimen table which was used to provide incentive to reduce wastage of leather from which shapes were being cut for wrist watch straps is given on page 256.

The attainable target set for this was a rate of £0·46 for a 100 performance and 85 per cent material usage. Earnings varied according to performance and usage. In practice, the accent is placed on one or the other according to the relative cost of labour and material.

Performance	Per cent material usage					
	95	90	85	80	75	70
120	£0·60	£0·58	£0·56	£0·54	£0·52	£0·50
110	£0·55	£0·53	£0·51	£0·49	£0·47	£0·45
100	£0·50	£0·48	£0·46	£0·44	£0·42	£0·40
90	£0·45	£0·43	£0·44	£0·40	£0·40	£0·40
80	£0·40	£0·40	£0·40	£0·40	£0·40	£0·40
70	£0·40	£0·40	£0·40	£0·40	£0·40	£0·40

(iv) Incentives for inspection

Incentives can be designed for inspection if careful consideration is first given to their duties, and the incentive built around these. One example of a successful incentive was applied to inspections of pressure gauges. These had to be 100 per cent tested for accuracy, and every faulty one rejected—without exception. Payment was made for each one tested based on measured work values, using the standard time system described in (*a*) (ii), but for every faulty one found, the credit given (and therefore the bonus) was doubled. A re-inspector (who was a chargehand) inspected all rejected gauges and took random checks on all those passed. If either faulty gauges were found amongst those passed as satisfactory, or satisfactory gauges found amongst those rejected, the bonus of the inspector responsible was heavily reduced.

Incentives for indirect labour

Payment by results systems can be applied also to indirect labour (i.e. labour which contributes indirectly to production, as labourers, storemen, maintenance workers, toolmakers, etc.). They can be classified into two main groups—

(*a*) Directly based output incentives. These are usually based on measured or estimated work standards.

(*b*) Indirectly based output incentives—which are those where payment depends on the performance of the direct workers.

(a) Directly based output incentives

Directly based incentives for indirect workers can take the form of any of those described for direct labour, provided that the work can be measured to within an accuracy of $\pm 2\frac{1}{2}$ per cent. Due to the nature of most work done by indirects, however, accuracy of this order is often not possible, and therefore one of the differential or stabilized output incentives described under sections (b) and (c) of 'DIRECT LABOUR' will be more suitable.

(b) Indirectly based output incentives

One example of an indirect worker on which such an incentive could be applied is a service hand, i.e. one whose duty it is to feed direct operatives with work. As the direct operatives increase their output, the work the service hand has to do will increase in proportion, and the earnings can be made to rise at the same rate.

Incentives of this type are never so successful, being at best a compromise. It is far better to measure the work which the indirect worker does, and to base his earnings on this.

Incentives for supervision

The introduction of an incentive scheme and labour cost control may change the character of the duties of supervision, the accent being on the reduction of waiting time and similar losses, in order to sustain output. In addition to this, if there is a production control system in operation in which work is programmed to the department at periodic intervals of a week or a month, it may be desirable to direct the effort of the foreman to prompt clearance of these programmes in chronological order.

If, therefore, the intention is to run the department by keeping waiting time and overtime to a minimum, and to clear the programmes of work promptly, then payment could be dependent on the performance of the foreman in carrying out these. Clearance of programmes is calculated by expressing the number of batches of work cleared as a percentage of the number originally programmed, irrespective of their size. All arrears are brought forward to next week's programme, but count as if the number of batches had increased by the number of weeks in arrears. For example, if two batches were not completed during one week, these would count as two batches in arrears. If, however, the same two batches were still uncompleted at the end of the second week, these would count as four batches in arrears, and so on. This ensures that all work is given equal importance and the incentive is to clear the work in chronological order as promptly as possible.

Overtime hours are added to hours waiting time to give "total lost hours".

257

The attainable *target* is 95 per cent clearance of programme with 20 lost hours.

Lost hours	Per cent clearance of programme					
	100	95	90	85	80	75
Nil	£7	£6	£5	£4	£3	£2
10 or less	£6	£5	£4	£3	£2	£1
20 or less	£5	£4	£3	£2	£1	—
30 or less	£4	£3	£2	£1	—	—
40 or less	£3	£2	£1	—	—	—
50 or less	£2	£1	—	—	—	—

Values shown are weekly bonuses. Obviously, different scales and values would be constructed to meet particular circumstances.

Learning allowances
Before applying an incentive to an operator who is strange to it, or to the work, it is often beneficial to arrange that their bonus is paid on an ascending scale of values. There are several ways of doing this, depending on the type of incentive, but the general principle is that an allowance is credited to the operators—usually up to a maximum of standard performance—and this allowance is gradually reduced over a period of time ranging from 3 or 4 weeks to 3 months or more according to the skill factor necessary to do the work. For most applications, however, 4 weeks is sufficient and is commonly used in practice.

Two ways of applying this to output incentives are given below—

(*a*) 1st method. Allowance on standard hours earned. Bonus is paid according to the following scale, but only up to a maximum of standard performance during the learning period.

   (i) 1st week. Add up to 30 per cent extra to the standard hours earned.

(ii) 2nd week. Add up to 20 per cent extra to the standard hours earned.

(iii) 3rd week. Add up to 10 per cent extra to the standard hours earned.

(iv) 4th week. Add up to 5 per cent extra to the standard hours earned.

Thereafter, no allowance would be given, but the bonus could be above standard if this is genuinely earned.

(*b*) 2nd method. Allowance on operator performances. Bonus is paid according to the following scale with a maximum of standard performance during the learning period.

(i) 1st week. Add up to 30 points to the operator performance index.

(ii) 2nd week. Add up to 20 points to the operator performance index.

(iii) 3rd week. Add up to 10 points to the operator performance index.

(iv) 4th week. Add up to 5 points to the operator performance index.

As before, no allowances would be given thereafter, but bonus would be paid according to the actual performance index earned.

Different scales can, of course, be constructed to suit different conditions.

The choice of incentive system

The type of incentive system most suitable for any particular application will be generally influenced by the following circumstances—

(*a*) Whether the work can be accurately measured. Some form of differential or stabilized output incentives may have to be considered in those situations where it is neither economical nor practical to measure the work accurately.

(*b*) Whether the work is process-controlled or not. These may need specially designed schemes.

(*c*) Whether the emphasis is on high output, high quality, material conservation or other considerations.

Whatever the circumstances, however, it is vital that the incentive directs effort to the desired end and that it is—

(*a*) Simple to operate and understand.
(*b*) Foolproof and cannot easily be manipulated.
(*c*) Fair, rewards being in proportion to effort.

(*d*) Not a cause of excessive fluctuations in pay, where these are not accompanied by equally deliberate fluctuations in effort.

(*e*) So designed that the targets set are regularly attainable by an average worker.

Although piecework or the standard time system output incentives are amongst the fairest, they are not suitable in those situations where management cannot guarantee an even flow of work, or where the standard times cannot be set to within fairly accurate limits. Amongst the most popular are the stabilized output incentives because they reward individuals proportionate to effort, but do not cause large fluctuations in pay.

In the final analysis, it must be remembered that the success of an incentive lies as much with its careful administration and with good industrial relations as with its actual form.

Launching an incentive

The success of an incentive will also depend as much on the way it is launched as on its design. It is therefore most important that the terms of reference of the incentive agreement are fully discussed and drawn up with the workers' representatives prior to its introduction, and that supervision are kept fully informed throughout.

A suggested procedure is given as follows—

(*a*) Obtain a reference period and calculate the amount of money which could be offered as an incentive payment at the standard rate of performance (see Chapter 17).

(*b*) Draw up draft terms of reference for the incentive. Include in this a learning period to give encouragement to the workers during the early stages.

(*c*) Draw up revised terms of reference after discussion with management as necessary.

(*d*) Management offer the scheme to the workers concerned, preferably via the shop steward or other workers' representatives or with them present. Both supervision and work study personnel should also be represented.

(*e*) Design the work sheets and draw up the final terms of reference for the scheme.

(*f*) Launch the incentive, spending as much time as possible helping the work people affected to understand it, to book their time correctly, and to achieve satisfactory bonus earnings at the earliest possible time.

Particular attention should be given to the correct booking of waiting time and other lost time and excess work allowances, as often operators will not realize the difference that this makes to the bonus earnings. It is also important that this time is properly classified after booking, as it will form the basis of labour cost control.

Revisions of incentive conditions

An incentive that is suitable for conditions at a particular point in time is not necessarily suitable for circumstances in the future. Incentives also tend to deteriorate over a period due to difficulties in administration, pressure from the shop floor, or poor supervision.

It is therefore recommended that any incentive is only applied for an agreed period of say, three years, after which negotiations are conducted for its complete revision based on the experience gained in running it. Such a revision might well coincide with a new wage agreement.

15

Labour cost control

One of the most important benefits to be derived from a work measurement application is the establishment of a known level of output and cost. For if a department functions on a well-designed and properly administered incentive based on measured work values, it will tend to produce at optimum rate with constant labour cost if kept adequately fed with work. If also requirements of the bonus system are such that all production losses in the form of waiting time and excess work are recorded, deviations from the standard rate of working can be collated and presented regularly to management to give clear indication of the nature and extent of these, so that appropriate action can be taken to reduce or eliminate them.

Before the principle of this system can be fully explained, however, it is necessary to understand clearly some of the terminology used.

Direct work
This is work performed on a saleable item and which directly changes its shape, form, standard of finish or classification. Press operations change the shape of a product by blanking out pre-formed shapes or piercing holes, etc., assembly operations change the form of a product by assembling various pieces, and polishing changes the standard of finish, etc.

Tool making is not direct work with respect to the products made by the tool, but can be classified as such with respect to the tool itself, since this is a saleable item taken on its own merit.

Measured work
Work is said to be measured if it can be expressed in units of standard or allowed time. The term is usually applied to describe useful or productive work.

Uncontrolled work
This is useful work which cannot be expressed in units of standard or allowed time because it has not been measured.

Ancillary work
This is work which closely contributes to the performance of direct work, but which is subservient to it. It would include preparation work, changing over and setting up of machines, etc.

Excess work
This is work done extra to that planned or allowed for, but necessary due to some fault in manufacture or administration. It can either be measured or uncontrolled.

Waiting time
Waiting time is unproductive time caused by operators being prevented from working by some administrative fault. In this category is time lost due to insufficient work flow, machine or tool breakdown, etc.

Allowances
In certain cases, allowances may have to be made to operators during the administration of a bonus system or other reasons. These could include learners' allowances, which enable untutored labour to earn at incentive rates until proficient in the work (see Chapter 14); or policy allowances, which are allowances given to overcome certain difficulties in producing and earning target wages which are outside the control of the operators. Policy allowances should not be included in standard times as they destroy the conception of measured work values.

Operator performance
Operator performance is an indication of the effectiveness of a worker or a group of workers whilst on measured or estimated work. If—

t = total standard times earned
T = attendance time
l = total waiting time or other lost time
e = total time in excess work or other allowances
u = total time on uncontrolled or unmeasured work

Then—

$$\text{Operator performance } OP = \frac{100t}{T - (l + e + u)}$$

Department performance

This is an indication of the effectiveness of a department or section in producing work.

If—

p = the assessed performance at which uncontrolled (unmeasured) work is produced.

Then—

$$\text{Department performance } DP = \frac{100t + up}{T}$$

For example, if a department produced 300 standard hours and 100 uncontrolled (unmeasured) hours of work in 450 actual hours, the uncontrolled work being assessed at a performance p of 60 then—

$$DP = \frac{(100 \times 300) + (100 \times 60)}{450}$$

$$= 80 \text{ BSI}$$

Sometimes, it may be decided that certain waiting time or diverted (i.e. lost) time is not the responsibility of the department, and that certain hours have been allocated for the operators to carry out certain other work, e.g. cleaning machines, etc., which is not accurately measured. In these cases, the department performance is obtained from

$$DP = \frac{100t + up}{T - (A + L)}$$

where A = the allocated work (cleaning, etc.)
L = the waiting time and diverted time for which the department is not responsible.

Overall performance

This is an indication of the net utilization of labour in producing useful work.

$$\text{Overall performance } (Op) = \frac{100t + up}{T - A}$$

Performance and rating

The difference between performance and rating can be explained as the rate of producing work units and the rate of producing basic times respectively. In this context work units means standard times plus uncontrolled time at assessed performance. If, therefore, an operator takes exactly the relaxation allowed in the standard times he is working on, his average rating and operator performances will be identical.

Attendance time

This is the time spent by a worker at the place or places of employment, whether working or not. It is often referred to as CLOCK HOURS also.

Standard costs

A standard cost is the cost of something at an agreed standard. Standard costs can therefore be applied to both direct and indirect measured work units on different cost basis, e.g.

(*a*) Direct standard cost per direct standard hour (or minute, etc.).

$$S = \frac{D}{w}$$

where D = the gross wages which would be paid to direct workers at standard performance in a standard working week.

w = the number of work units produced by the same direct workers in a standard working week.

Therefore, if a group of 10 direct workers are paid gross wages of £200 when working at standard performance in a standard working week of 40 hours, since the number of standard hours which would be produced under these conditions would be $40 \times 10 = 400$, then the direct standard cost per direct standard hour is—

$$S = \frac{£200}{400}$$

$$= £0.5 \text{ per direct standard hour}$$

(*b*) Indirect standard costs can either be expressed per indirect standard time or per direct standard time. In the former case it is calculated similarly to (*a*) above, but in the latter case it is—

Indirect standard cost per direct standard hour (minute, etc.).

$$S = \frac{d}{w}$$

265

Here d = the gross wages which would be paid to indirect
workers at standard performance of direct workers in a
standard working week.

For example, if in case (*a*) there were two indirect workers in the
department (say labourers) and they would receive £20 per week
each for a standard working week, then the indirect standard cost
per standard hour is—

$$S = \frac{£40}{400}$$

$$= £0\cdot1 \text{ per direct standard hour}$$

(*c*) Total standard cost per direct standard hour (minutes, etc.)
would be—

$$C = S + S$$

In this example this would be,

$$C = £0\cdot5 + £0\cdot1$$
$$= £0\cdot6 \text{ per direct standard hour}$$

Actual costs
If standard performance is not achieved in some way, there may be a
difference between the standard and the actual costs. Actual costs
are—

$$A = \frac{G}{W + up}$$

Where G = gross wages paid out to a group of workers.
W = the total work units produced.
u = the uncontrolled (unmeasured) work produced.
p = the assessed performance at which uncontrolled work
is produced.

Excess costs
Excess costs are the difference between the actual and standard
costs—

Excess cost per standard hour = (Actual cost per standard hour)—
(standard cost per standard hour)
and
   Total excess cost = (excess cost per standard hour) × (standard
hours produced).

Excess costs are made up of losses due to—

(*a*) Waiting time and other lost time.
(*b*) Excess work.
(*c*) Special allowances.
(*d*) Make-up.
(*e*) Overtime premium payments.
(*f*) Indirect excess cost due to a surplus of indirects over a standard complement.

If the waiting time, other lost time and excess work booked on operator's work sheets is classified, this can form the basis for an analysis of excess items which is shown on a Control Sheet.

Suggested analyses of such items are listed below—

  (i) Wait equipment.
 (ii) Wait instruction.
(iii) Machine not available.
 (iv) Machine breakdown.
  (v) Wait work.
 (vi) Discussions with supervision.
(vii) Wait material.
(viii) Trade union negotiations.
 (ix) Medical treatment, etc.

Make-up
This is described in BSB138:1969 as "the amount of adjustment in terms of money or time required to bring a worker's earnings up to his guaranteed minimum".

It can be explained as follows—

Assume that an incentive is operating where bonus is paid for all operator performances over 75 BSI, and that the rate of pay at 75 BSI is 37·5p per hour and at 100 BSI 50p per hour, but that the 37·5p per hour payment is minimum.

Now consider the case of an operator whose performance is 70 only. He will be paid at 37·5p per hour, but he will only have earned 35p per hour, or his *make-up* will have been—

75–70   = 5 standard hours per attendance hour, i.e. 40 × 5 = 200 standard hours for a 40 hour week.
37·5p–35p = 2·5p per attendance hour or 40 × 2·5p = £1 for a 40 hour week.

Make-up is a source of loss and therefore is an *excess cost*.

Item	Where used as a control
Measured work	Should not be shown on a weekly control sheet. It is used as a basis for control in calculating performances and ratios of uncontrolled and ancillary work, and actual costs
Uncontrolled or unmeasured work	Can be shown on weekly control sheet for supervision as an amount, and on weekly control sheet for management as a percentage of measured work
Ancillary work	Analysed over a period, can form a useful basis for costing. Not usually used for routine control
Attendance time	This is best expressed as the equivalent number of employees for manning control. The ratio is obtained by dividing the attendance time by the number of hours in a standard week
Operator performance	Use on weekly control sheets for management and supervision
Overall performance	Can be used as an alternative to operator performance
Department performance	Use on weekly control sheets for management and supervision
Standard costs	Used as a basis for control and should not be shown on a weekly control sheet
Actual costs	Used to calculate excess costs and should not be on control sheet. Can form a useful basis for costing when analysed over a period
Excess costs Waiting time Excess work Special allowances Make-up Overtime premium Indirect excess	Excess costs are best shown expressed as a percentage of the standard cost per unit of time and entered on a management control sheet. Total and individual excess cost should be expressed in time units on a supervision control sheet. Analysis of the excess items either in time or money is usually required by exception by management, i.e. when total excess costs are running high

Fig 67 Summary of labour cost control information indicating where it should be used. Control sheets for management and supervision should be designed to present the information weekly.

Overtime premium

Overtime is that part of attendance time which is over the standard working week, and is usually paid at a higher rate. For example, if an operator is in attendance for 50 actual hours when he is required normally to work only 40 hours then he is said to have worked 10 hours overtime. If this 10 hours is made up of—

5 hours on Saturday paid at time-and-a-half
5 hours on Sunday paid at double time
then he will receive an extra—
$(5 \times \frac{1}{2}) + (5 \times 1)$
$= 7\frac{1}{2}$ hours pay

This pay will be made at an agreed rate and is known as overtime premium payment. It is an *excess cost*.

Control sheets

Labour cost control is operated by publishing regular returns of the conditions of labour costs and performances in departments as a form of management control. The returns are not accurate balance sheets but are intended as a guide to management where remedial action should be taken to control costs to as new a standard level as possible. It is important to realize that they do not in themselves have any other function than to present information. If action is not taken when losses are clearly presented, the situation will tend to worsen and the benefits obtained from work study eroded away.

The chart, Fig 67, gives an indication of what information could be presented at different levels of supervision and management. Control sheets should be constructed and operated with an agreed distribution based on this model.

The analysis of excess items can be expressed either in actual time or in approximate cash value. When presented to supervision it is more effective if such analyses are expressed in hours, as they mean more to this level of management in this form.

They can be converted to an approximate monetary value by dividing the total lost and excess work time by the total excess cost and multiplying this rate by each individual excess item. This is a good enough approximation for control purposes, and no useful purpose is served by attempting any greater accuracy.

16

Overhead and material costs

From the very nature of work study and the effect that its application can have on cost, it is important that all who practise it have a good general appreciation of cost accounting methods.

Fundamentally, cost is evaluated in terms of human effort. Material in its basic form which is lying in its originating source costs nothing. Iron ore only begins to have a value once it has been mined and made available for other uses. Water falls freely from the atmosphere in the form of rain, and reappears just as freely in springs, rivers and lakes. But immediately it is filtered and piped to a consumer's premises it acquires a cost equivalent to the human effort expended in so doing.

Material cost can again be raised by further human energy. To convert iron ore to iron and steel requires a great deal of effort, from the complex administering and technical skills during manufacture, to the design and running of the special plant, as well as the sheer manual force necessary to carry out the process.

This is why the cost of a product or service is usually regarded as being made up of three elements—

(a) Labour, which is an evaluation of direct human effort.

(b) Material, which is an evaluation of that human effort expended in basic raw materials in its natural source in order to make it available where it is wanted in a more convenient form.

(c) Overheads, which is an evaluation of human endeavour in co-ordinating the direct and indirect human effort necessary to bring about the final product or service.

Cost, to a high degree, will determine price; but price is also subject to fluctuation according to supply and demand. Thus a rare product in high demand will raise the price, or in other words, inflate the cost of the labour used in producing it. Conversely, a product in continuous but not high demand may fix its own price by reason of the level of supply. If such a product can be obtained from different sources all of which produce it for different amounts of effort, then the price per human effort will also fluctuate. In other words, if the price is constant, the organization producing by the most efficient means will be able to give more per unit effort, i.e. pay higher wages or make more profit.

The value of work study is therefore again demonstrated, since it seeks to produce the same result with less effort, thus bringing more prosperity.

Labour cost

As has been previously explained, total labour cost can be considered as consisting of both direct and indirect components. It is most accurately determined by using standard times based on work measurement and labour rates which are inclusive of allowances for tolerable amounts of excess cost.

Labour cost can be regarded as being *variable* with respect to output—i.e. the greater the output, the higher will be the labour cost in direct proportion.

Material cost

Material cost is the cost of all the direct material used in manufacture of the finished product, and should be inclusive of normal wastage in the form of scrap. The cost of paint, metal deposits, or other finishes can either be classed a direct material and included in this cost or, if it is a very small amount, included in the overhead cost.

Overhead costs

Overhead costs are a form of indirect cost not readily associable with direct activity. Typical examples of such costs are—

(*a*) Cost of factory premises   (rent, rates, repairs to buildings etc.)

(*b*) Directors and executives   (directors' fees, directors' and executives' salaries, bonuses, etc.)

(*c*) Heating and lighting costs   (heating and lighting costs of factory premises and offices)

271

(*d*) Plant depreciation   (depreciation of machinery and plant)

(*e*) Electricity   (power used for machinery and plant only)

(*f*) Other services   (water and gas used for industrial processes)

(*g*) National Health contributions   (for direct employees)

(*h*) Staff salaries   (junior management and office staff)

(*j*) Sales and marketing expenses   (sales representatives, agents fees, advertising, etc.)

(*k*) Plant maintenance   (maintenance staff, spares, oil, grease, etc.)

(*l*) Canteen subsidy

(*m*) Selective employment tax

(*n*) Sundry costs   (telephone, stationery, petty cash, etc.)

Classification of overhead cost

Examination of overhead cost will reveal differences in their nature. Items (*a*) to (*d*) would be expected to remain constant, irrespective of the output; these are known as "fixed" overheads. Items (*e*) to (*g*) would most likely vary directly with output; whilst items (*h*) to (*n*), although tending to vary with quantities produced, would do so to a much less marked degree than items (*e*) to (*g*). These are known as variable and semi-variable overheads respectively.

Careful assessment of each cost will also enable bases for their variation to be established. For example, the consumption of electrical power will vary directly with machine hours; whereas National Health contributions for direct workers will vary according to the direct labour employed, i.e. to the output in standard hours of work.

These costs can also be allocated departmentally. Depreciation can be distributed according to the capital of the plant installed and used in the department, electrical power costs divided according to kilowatt-hour usage—or better still separately metered for each area in which it is used, etc. Semi-variable overheads can usually be resolved into fixed and variable categories within themselves by further analysis—similar to the analysis of work elements during work measurement.

Allocation of overhead cost

One of the most common methods of determining overhead rates for allocation to the product is by constructing *break-even charts*. An example is shown in Fig 68. The base line of the graph is established in units of production per annum, and the vertical ordinate in £'s expenditure. The value of the "fixed" overheads is next drawn as a line parallel to the base. To this is successively added the variable overheads, the labour cost and the material cost from previously determined values at known levels of output to obtain a total expenditure line. The income line passing through the origin is

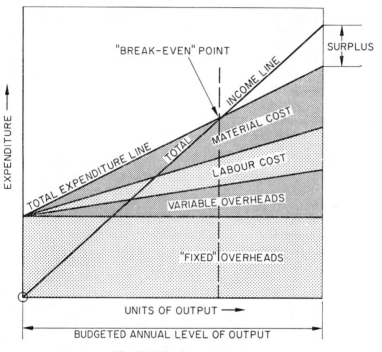

Fig 68 "Break-even" chart

based on a realistic selling price per unit to give a predetermined surplus (or net profit). At a certain point the total expenditure is known as the "break-even point" since at this point in time and output both accounts will be the same. Beyond this there accumulates a surplus which can be appropriated on a profit and loss account.

On the chart shown, the total overhead rate per unit will be the total overheads divided by the output at break-even point. In practice the details are often more complex, several such charts having to be constructed for different departments and bases of variation. The principle, however, remains exactly the same.

273

Marginal cost

The *marginal cost* is the total cost of all the variable elements of cost. In other words, it is the total cost less the fixed overheads content.

Costing methods

There are several different types of costing methods which are used for different applications. Some of these are briefly described below.

(a) *Job costing.* This is a method of costing where each particular job is regarded as a separate cost unit, apportionment of the elements of cost being directly applied in the most accurate way possible.

(b) *Contract costing.* This is exactly the same as job costing but has a different terminology when applied to large-scale contracts.

(c) *Output costing.* This method of costing can be used in breweries, dairies, chemical works, etc. where relatively few products are made and sold by the pint, gallon, etc. and the cost is required on this basis.

(d) *Process costing.* This is an elaboration of output costing, being a further breakdown into the cost of each process involved (e.g. the cost of brewing, bottling and other processes separately in a brewery).

(e) *Department costing.* This is separate costing of departments in an organization, e.g. a tool room or maintenance department could be separately costed and controlled.

(f) *Operation costing.* This is used when it is necessary to cost a service as opposed to a product (e.g. transport per mile, gas per cu. ft in a gas works, or the cost of work study, etc.).

(g) *Standard costing.* This is costing based on averages set as standards. Standard labour costs are determined from standard output levels and standard rates of pay and hours of work, and labour cost control can reveal and regulate any deviations from them. Material is also costed according to standard usage, and overheads from budgeted levels of output and expenditure.

Other terms used

Some other terms used in costing are useful to know, such as—

(a) *Prime cost.* This is the cost of direct materials, direct labour and current departmental expenses (or departmental overheads).

(b) *Factory cost.* This is the prime cost plus the total factory expenses (fixed and variable factory overheads).

Control of expenditure of these accounts is carried out by Budgetary Control.

Fixed overhead recovery

Considering the previous example, let it be assumed that of the total overhead content of £4·575, £1·5 was "fixed overhead".

If by the application of work study, the output is raised from 200 units per week to 220 units per week with no increases in either material, labour or overhead rates, then the effect would be as follows—

Recovery of fixed overheads before application of work study

$$= 200 \times £1·5 = £300 \text{ per week}$$

Recovery of fixed overheads after application of work study

$$= 250 \times £1·5 = £375 \text{ per week}$$

Therefore there will be an *over-recovery* of overheads of £75 per week, or in other terms, an increase in revenue of—

$$48 \times £75$$
$$= £3,600 \text{ per annum}$$

This consideration is particularly important when dealing with cost reduction of process-controlled work. In many cases the depreciation is so high that more savings can often be made by *increasing* the labour cost if the machine utilization (and therefore the output) can be raised by so doing (see Example in Chapter 10), and the increased recovery of fixed overheads exceeds the losses in labour cost.

17

Cost reduction reports

Reference periods

Since work study is a cost reduction technique, it is essential to be able to show positive results from its application. For this purpose a truly representative period must be selected before any changes take place, and the conditions within it carefully recorded. Subsequent reference can then be made to this when calculating and reporting on subsequent benefits.

Such a reference period must be indisputably authoritative, otherwise any claims based on it can be challenged. Data used must be obtained from reliable sources, if possible, supported by the official documentation of the organization and verified by responsible members of the management team. Unusual circumstances should be avoided, such as a short holiday interposing, or occasions when reorganizations are taking place. It is also important that the reference is over a sufficiently long period of time. Usually a month is sufficient, unless there are seasonal or other fluctuations present, or there is a large variety of products, when a longer interval may be necessary.

The form of the reference will vary according to circumstances. Some of these are now considered in detail.

Direct labour references

This is a record of the performance, output and cost of direct labour. It can either be used in preparation for introducing work measurement based incentive bonus systems, or before method

278

changes take place—some of which may include the installation of new plant.

The procedure for obtaining a reference of this type is as follows—

(*a*) Obtain the output over the period.

(*b*) Obtain the gross direct hours and the gross direct payroll.

(*c*) Calculate the output in standard time.

(*d*) Calculate the reference period department performance.

(*e*) Calculate the reference period direct cost.

In order to aid explanation, these steps are now examined more closely.

(*a*) *Obtain the output over the period*

This should be extracted from official production records or, if this is not possible, a record should be temporarily introduced specifically to collect this information. Details will be required of the quantities produced (or work in other forms), work preparation and machine setting details, etc. Where more than one operation is performed on the same part or piece of equipment, details of this will also be required.

Estimates of output should never be used, as these are rarely acceptable.

(*b*) *Obtain the gross direct hours and the gross direct payroll*

This is the gross hours worked and the gross wages paid to all the direct workers reponsible for the output recorded above, and over the same period of time. It should be obtained from the wages department.

(*c*) *Calculate the output in standard time*

This is carried out by multiplying the output by the standard times and totalling the results. For example, output details may have been recorded as follows, standard times being obtained by work measurement—

400 Item A @ 15 standard hours per 100	= 60 standard hours
250 Item B @ 30 standard hours per 100	= 75 standard hours
600 Item C @ 20 standard hours per 100	= 120 standard hours
10 settings of machine @ 1 Standard hour each =	10 standard hours
Total (per week)	= 265 standard hours

(*d*) *Calculate the reference period department performance*

This is obtained by dividing the total standard hours by the gross direct hours. If there has been any work included which has not been measured, this is deducted from the gross direct hours.

Cost reduction reports

For example, if the gross direct hours are 403, and the amount of work not measured is 3 hours, then—

$$P = \frac{100W}{T - U}$$

where P = department performance
W = the output in standard time
T = the gross direct hours
U = the amount of hours of work not measured

or $P = \dfrac{100 \times 265}{403 - 3}$

   = 66·3 performance

(e) Calculate the reference period direct cost
This is obtained by dividing the gross payroll by the total standard time, plus any work not measured at reference period performance.

For example, if the gross payroll of a department was £106·8 per week, then the reference period direct cost would be—

$$= \frac{£106·8}{265 + 2}$$

= £0·4 per standard hour

The extra hours added to the total standard hours of 265 is calculated from—

$$\frac{66·3}{100} \times 3$$

= 1·989 (say, 2 hours)

Reference periods inclusive of indirect labour
This will be in a similar form to that described for direct labour, except that any indirect costs allied to the department are included.

For example, if there was a labourer and a storeman in the above department, whose wages together totalled £53·4 per week, the indirect cost per *direct* standard hour would be—

$$= \frac{£53·4}{267}$$

= £0·2

and the total cost—

$$= £0·4 + £0·2$$
$$= £0·6 \text{ total cost per direct}$$
$$\text{standard hour}$$

Reference period inclusive of overheads

The results of work study may allow the overall output to be increased. If this is so, there will be an added reduction in cost due to a greater rate in the recovery of fixed overheads. For example, if the number of units being produced is x, and the "fixed" overhead is F, then the fixed overhead rate per unit will be F/x. If by improvement in output, x is increased to $x + p$, then there will be an increased rate of recovery of "fixed" overheads equivalent to—

$$= \frac{Fp}{x}$$

To explain this more clearly by quoting a numerical example, if the standard hours representing the total output during the reference period was 1,000, and the "fixed" overhead was £1,000 over the same period, then the fixed overhead rate would be—

$$= \frac{F}{x}$$
$$= \frac{£1,000}{1,000}$$
$$= £1 \text{ per standard hour}$$

If the number of standard hours output can be increased by 100 to 1,100 by applying work study without increasing the "fixed" overheads, then assuming the selling price of the article remains constant, there will be an increase in recovery of overheads equivalent to—

$$= \frac{Fp}{x}$$
$$= £1 \times 100$$
$$\text{or } £100$$

This will either enable selling prices to be reduced or the revenue to be increased, whichever is considered to be the more advantageous to the organization.

Machine utilization reference periods

Increase in machine utilization should allow more to be produced, and therefore if the output rises, the benefits to be gained can be measured in terms of increased recovery of fixed overheads—particularly the depreciation rate of the machine.

Reference periods used for method study applications
The effect of method study which is successfully applied should either be to reduce the labour cost, increase the output, increase the machine utilization, or a combination of all or some of these.

Revised standard time values as a result of work study should be used to obtain labour cost reference as explained earlier, and the current fixed overhead rates then calculated as a basis for increased overhead recovery, if increases in output will result.

References for material utilization
Sometimes, benefits are apparent from work study due to better material utilization. This could be either as a result of method improvement or from the application of bonus incentive designed to stimulate higher quality standards and more care in the use of raw material.

Reference periods of this type should record the amount of material used, set against the eventual output. To obtain a true value of this, the amount of stock and work-in-progress of both material and finished goods must be taken into account when calculating the usage per unit of output.

Surveys
The presence of waste can be detected in various ways. Some of these are—

(a) By direct observation of conditions on the shop floor or in the offices.
(b) By consideration of the sales techniques in use.
(c) By investigation into the methods of production and stock control.
(d) By examination of the company balance sheet.
(e) By examination of product design.

Such investigations are known as *Surveys*. They should only be undertaken by the most experienced observers as a great deal of skill and judgement is required.

Where the use of work study is contemplated the procedure is as follows—

(a) Discuss the investigation with shop supervision and trade union representatives.
(b) Assess the department productivity of the area under investigation by observation of the general pace of working

and incidence of delay. Choose a typical period which represents average conditions and err on the generous side when making the assessment. Performance levels where there is no incentive payment under average supervision will rarely be more than 65 BSI. Under a poorly designed and administered incentive it will often be around 80 BSI.

(c) Obtain details of all names, clock numbers, labour grades, hours of work and gross wages of all personnel in the area from the wages office over a period of at least one month.

(d) Obtain the department output and the fixed overhead rates for the same period if reliable records of these are available.

(e) Record any other relevant information which may be contributing to low productivity such as working environment, congestion, poor payment methods, employee morale, condition of equipment, etc.

(f) Estimate the future department performance after application of work study techniques. For example, if work measurement based incentive bonus is to be used, it may be assumed that eventually, operators will work at standard (100) performance, but when allowance has been made for inevitable excess costs, department performance would be around 95.

(g) Calculate the possible percentage increase in performance P from—

$$P = \frac{100(fp - sp)}{sp}$$

Where sp = assessed survey performance
fp = finally assessed department performance after work study application

For example, if the survey performance is 65, and the final performance is assessed at 95 then—

$$\text{Percentage increase in performance} = \frac{100(95 - 65)}{65}$$

$$= 46\%$$

(h) Assess the possible increases which will have to be paid out in order to achieve the results of work study. If an incentive is contemplated, benefits could be shared, half going to management and half to labour.

For example, if the increase in performance is estimated to be 46 per cent, the incentive offered to labour would be in the region of an 18 per cent increase in earnings, as this shares the saving equally between management and labour, according to the following formula—

If P = the assessed percentage increase in performance
E = the possible percentage increase in earnings

Then $R = 100 \left(1 - \dfrac{(100 + E)}{100 + P}\right)$

Where R = the percentage reduction in cost at the current output level

Applying this to the example, if $E = 18\%$ and $P = 46\%$ then

$$R = 100 \left(1 - \frac{100 + E}{100 + P}\right)$$

$$= 100 \left(1 - \frac{118}{146}\right)$$

$$= 19\%$$

which is nearly the same as the increase in earnings. To arrive at the halfway unit in this way, construct a graph similar to that shown in Fig 69 (p. 291).

(*j*) Assess the possible reduction in cost which should result from applying work study or other techniques. The procedure from this point onwards is as follows—

(i) Calculate the expected cost reduction at current output. In the example this would be—

$$\frac{R \times \text{Gross payroll}}{100}$$

Assuming for the example, a present gross payroll of £500 per week—

$$= \frac{19 \times 500}{100}$$

$$= £95 \text{ per week}$$

(ii) Calculate the expected cost reduction at expected output levels.
In the example, this would be—

$$\begin{array}{c}\text{(Expected cost reduction at} \\ \text{current output)}\end{array} \times \frac{100 + i}{100}$$

where i = the expected percentage increase in output

If there is to be expected increase in output of (say) 40 per cent, then the cost reduction would be—

$$£95 \times \frac{140}{100}$$

$$= £133 \text{ per week}$$

(iii) Calculate the net benefits by deducting from the cost reduction previously calculated any extra expenses such as cost of bonus clerks, special tools, special plant, etc.

The above calculations apply to reductions in labour cost only. Where there will be extra benefits in the form of increased recovery of fixed overheads, these should also be calculated at this stage.

(*k*) Assess the time and number of staff necessary to achieve the results. Some guidance is given on this at the end of Chapter 11.

(*l*) Arrange the results of the survey as a report to management.

Survey reports

Reports sent to management should be as short as possible. Few executives have either the time or the inclination to read lengthy reports. A manager is much more likely to be impressed by a few vital facts presented clearly and concisely. If it is still considered necessary to elaborate in order to substantiate any claims for results achievable or achieved, an appendix can be added by way of explanation. By this means, more detail can be introduced, whilst still presenting a short and quickly absorbed main report.

Headings should be typed in capitals starting flush with a left-hand (1 in.) filing margin and underlined—

(*a*) INTRODUCTION

A short explanatory statement is all that is needed under this heading, e.g.—

"This is a report on a survey carried out in the machine shop by members of the Work Study Department in order to investigate the scope for increasing output and reducing labour cost."

(*b*) GENERAL OBSERVATIONS

This should consist of a brief explanation of what took place during the survey together with general comments on the impressions formed.

"Several visits were paid to the area and observations made on the machining processes and on the general level of performance of the operators. In each case, the foreman,

Mr. J. Simmons was in attendance, and the shop stewards kept fully informed of the reason for the visits.

Some of the operating speeds of the machines appeared to be inconsistent, and a few checks on identical types and makes of capstan lathes machining similar materials found them to be running at different speeds. There was also a high incidence of waiting time which was felt to be due mainly to congestion on the shop floor, particularly in the main gangways, these being continually blocked by excessive amounts of work in progress.

From these observations the present department performance was assessed at 60 BSI.

"Details of wages for the month of September were obtained from the wages department. It was agreed by Mr. Sanders, the Chief Accountant that this was a typical period to choose as a reference. Examination of this payroll and comparison with other labour grades indicates that the present pay rates are generally lower than in other departments.

It is not therefore considered that inflationary wage rates would result if a bonus incentive system was introduced in the department which uplifts present rates to around 20% for a target level of performance. This view was confirmed by Mr. Hudson, the personnel manager."

(c) RECOMMENDATIONS

Again a brief statement—

"It is recommended that a method study exercise should be carried out in the department in order to rationalize the machine operating speeds, modify the workshop layout and relieve the congestion. This should then be followed by work measurement and the introduction of an incentive bonus system based on standard times."

(d) BENEFITS TO BE OBTAINED

A statement of fact substantiated by authenticated figures given in the appendix—

"It is estimated that these changes could be achieved by the Work Study Department in 18 weeks and would yield the following results—

(i) An increase in output from the department of at least 50 per cent.

(ii) For this output, it is estimated that an incentive would raise earnings to 20 per cent of their present levels. This is in line with earnings received by similar operators in other parts of the organization and would relieve the recent trade union pressure for wage increases.

(iii) A reduction of 20 per cent in the present labour cost of machining. Based on the present payroll of £930 per week, this would amount to savings of £280 per week or approximately £14,000 per annum at the enhanced level of output.

(iv) If the cost of the two work study analysts for 18 weeks is evaluated at £3,200 inclusive of department overheads, this represents an annual return in investment of over 400 per cent.

(v) It has been confirmed by the data processing department that the calculation of bonuses would not significantly increase administrative costs."

(*e*) APPENDIXES

Appendixes should contain complete details of names, clock numbers, hours worked, trades and occupations of all personnel in the department over the period mentioned, showing how the gross payroll has been obtained. Reference should be made to the authorities from which this information has been obtained (e.g. Mr. Jones, Wages Department, etc.).

Other details usefully included in the appendix could be the means by which the savings have been calculated. This would be based on the procedure given under the heading "SURVEYS" in this chapter. It is most important to set any benefits against involvement in capital or installation costs. Net savings are the true savings achieved.

Savings calculations

Financial and other benefits due to the introduction of work study methods can be calculated simply from reference period data, particularly if the work is being measured and the output properly recorded. Results which can be shown are as follows—

(*a*) Increases in output.
This can be calculated from—

$$I = \frac{100(CP - RP)}{CP}$$

where I = Percentage increase in output
   CP = Current department performance
   RP = Reference period department performance

For example, if the reference period performance was 60, and the current department performance (after application of work study techniques) was 95, then the increase in output would be—

$$I = \frac{100(95 - 60)}{60}$$

$$= 58\%$$

(b) Increases in earnings of operators.
This can be calculated from—

$$e = \frac{100(CE - RE)}{RE}$$

where e = percentage increase in earnings
CE = Current average hourly earnings
RE = Reference period average hourly earnings

Note The earnings levels should be calculated *net*, i.e. not inclusive of overtime premium, but inclusive of bonus payments.

For example, if the reference period earnings were £0·5 per hour, and the current earnings £0·6 per hour, then the increase in earnings would be—

$$e = \frac{100(£0·6 - £0·5)}{£0·5}$$

$$= 20\%$$

(c) The decrease in labour cost.
This can be calculated from—

$$L = \frac{100(AC - RC)}{RC}$$

where L = percentage decrease in labour cost
AC = present actual cost per standard hour (or minute, etc.) (obtained by dividing total standard hours output into the gross payroll for the department)
RC = reference period cost per standard hour (or minute, etc.)

For example, if the present cost is £0·6 per standard hour, and the reference period cost was £0·9 per standard hour, then the reduction in labour cost would be—

$$L = \frac{100(£0\cdot9 - £0\cdot6)}{£0\cdot6}$$
$$= 50\%$$

(d) The savings in labour cost.
This can be obtained from—

$$S = (AC - RC)P$$

where S = gross cost reduction in labour cost per week
         P = current output in standard time

For example, if the current weekly output is 700 standard hours, then the savings (or cost reduction) on labour cost would be—

$$S = (£0\cdot9 - £0\cdot6)700$$
$$= £210 \text{ per week}$$

This can be converted to the annual rate of savings by multiplying by the number of working weeks, e.g.

$$= 48 \times £210$$
$$= £10,080 \text{ per annum}$$

These savings should be expressed inclusive of indirect labour if relevant.

(e) The savings in fixed overhead recovery.
This can be obtained from—

$$S = \frac{F(100 + I)}{100}$$

where S = savings from over-recovery of fixed overhead
         F = total fixed overhead amount
         I = percentage increase in output

This can only be claimed if the output increased is actually sold, and the fixed overhead content is the whole of it allocated to the product under consideration.

For example, if by work study application, the output *and* sales of a company's sole product was increased by 10 per cent, and the fixed overhead was £20,000 per annum, then, the savings due to increased rate of recovery *if the selling price remains constant* is—

$$S = \frac{£50,000 \times (100 + 10)}{100}$$
$$= £5,500 \text{ per annum}$$

Cost reduction reports

Cost reduction reports should be sent regularly to management. This is an essential "window dressing" activity which will do a great deal of good in advertising the effectiveness and importance of work study—and the work study department—as part of the management team.

As with survey reports, these should be as brief as possible, and confined to relevant factual detail, being expanded in appendixes with references to the source of the information.

Suggested headings are as follows—

(*a*) INTRODUCTION

"This is a report on results achieved from improved floor layout and the introduction of a work measurement based bonus incentive in the machine shop."

(*b*) ALTERATION TO WORKSHOP LAYOUT

"Methods investigation commenced on the 5th October, and continued until the 23rd October, after which an improved workshop layout was drawn up and presented in a report. Suggestions were also incorporated in this report for the rationalization of machine speeds. The new layout was installed immediately afterwards, being completed by the 20th November."

(*c*) INCENTIVE SCHEME

"Meanwhile, work measurement was carried out and the labour was ready to be applied under incentive by the end of February of this year.

This bonus scheme has now been in operation for 6 weeks, and is running satisfactorily and the current department productivity is 90 BSI."

(*d*) RESULTS TO DATE

"The following current results are at present being achieved over the reference period established last September, these being based on official records from the wages department, the labour control records, and the production control departments.

Details of these together with the relevant calculations are given in the appendix—

Increase in output	= 58%
Increase in operations earnings	= 20%
Decrease in labour cost	= 50%

Gross savings = £210 per week
or approximately £10,000 per annum

Offset against this are installation costs of £3,900."

(*e*) CONCLUSION

"It is confidently expected that the output will further increase to over 60 per cent during the next 3 months. This should give net savings at the rate of £10,500 per annum.

In addition, there is a smoother flow of work through the department with consequently better housekeeping and better industrial relations due to the improved layout and the incentive scheme."

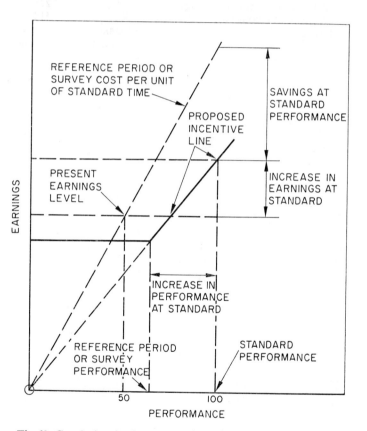

Fig 69 Graph showing how cost reduction can be shared between management and labour when applying a bonus incentive

(*f*) APPENDIXES

Detailed references to all figures obtained should be given in the appendixes, together with the actual calculations used to obtain the savings. Care must be taken to include any increased running costs, so that the claims made are *net*. It is also sometimes useful to compare results obtained with the survey forecasts, e.g.—

	Actual results	*Survey forecast*
Increase in output	58%	45%
Increase in operation earnings	20%	22%
Decrease in labour cost	50%	41%
Savings	£10,000 per annum	£8,900 per annum

18

Summary of procedures

Part 1—Introduction

Since work study is essentially a technique of cost reduction, the effectiveness of any application of it will be judged mainly on this basis. Each proposed investigation should therefore be carefully tested for economic validity before it is launched, for the cost could outweigh the achievable results—or the subsequent cost of maintaining the scheme could be so high that it would clog the work study department with routine work and prevent extension of operations into other areas urgently needing attention.

The choice of a particular set of techniques and the order in which they should be applied must always be based on economics. The attitude should remain severely practical. Theoretically, it is correct procedure to investigate and implement all method improvements down to the last detail before attempting to consolidate with work measurement and bonus incentive. But often this means prolonged periods of investigation and change when no tangible benefits will result. The waste will still continue while the investigation is on, and although this is to some extent inevitable, broad method study and simple changes combined with work measurement and bonus incentive will often bring about substantial cost reduction in a much shorter period. Further method improvements can then be made as a second stage while the profits from the first stage are still being felt. This should not normally be difficult if it is made clear that any future changes in method will mean changes in measured work values.

Where there is mutual trust and a sense of fairmindedness, it will be far less difficult to bring about change in any form. It is nevertheless essential to spend a good deal of time during preliminary investigations on the problem of the possible reactions of labour to the introduction of work study before any detailed work is commenced. If, for example, there is a basic objection to direct methods of measurement by the stop watch or other timing device, some indirect method such as a predetermined motion-time system, may be more acceptable. The techniques can be adapted to suit the conditions, but the target should still remain the same—the most effective cost reduction in the shortest possible time.

A work measurement programme should always be arranged to create data which can be built into work values with the absolute minimum need for time study in the future, and should in no circumstances be allowed to deteriorate into individual rate setting of operations of similar character. Moreover, the synthetic data should be so designed to enable the time values to be compiled quickly and with sufficient—but not over-sufficient—accuracy. For apart from the obvious economic reasons, few conditions are more demoralizing to a good work study man than to be fettered with routine work of this nature, instead of having free scope for his ideas on improving methods.

Another important fact to be remembered is that carefully planned work measurement combined with properly conceived bonus incentive will allow the introduction of labour cost control, which will then tend to reveal other faults in the organization, such as poor maintenance methods or lack of effective production control. The need for such changes cannot readily be seen by using method study alone, and there would almost certainly be great difficulty in assessing the financial benefits of so doing without such controls.

Work measurement applications using predetermined motion-time systems or broad method study techniques should follow exactly the same pattern as when using time study with a stop watch, except that the recording techniques and methods of obtaining elemental times will be slightly different. Adaptations of the procedure to suit these techniques are given in Part 5 of this chapter.

Part 2—The procedure in outline
The notes in parentheses after each stage indicate when action should be taken by work study staff alone, when by management alone, and when by a combined effort. Other notes in brackets show which parts of the procedure apply to work measurement only, which to team work measurement, and which to process-controlled work. To adapt the procedure for any application, these are either

included or omitted according to the techniques used. Examples are given in Part 5.

STAGE 1—SELECT AREA FOR SURVEYAL (combined work study staff and management action).

STAGE 2—SURVEY AREA OF OPERATION (work study staff).

STAGE 3—ASSESS POTENTIAL BENEFITS AGAINST IMPLEMENTA-TION COSTS (work study staff).

STAGE 4—REPORT ON RECOMMENDATIONS (work study staff).

STAGE 5—SANCTION INVESTIGATION (management action).

STAGE 6—CARRY OUT DETAILED INVESTIGATION (work study staff).

(a) *Further investigate prior to work study*
  (i) Discuss the investigation with supervision and the workers' representative.
  (ii) Select operators for preliminary study.
  (iii) Consult technical staff if the equipment or processes are not fully understood.
  (iv) Draw up preliminary work study programme.

(b) *Carry out initial work study observations*
  (i) Draw site plan or workplace layout.
  (ii) Take exploratory observations of present working methods.
  (iii) Report on any preliminary broad method changes needing management action and ensure their prompt installation.
  (iv) Plan main work study programme.

(c) *Carry out main work study observations*
  (i) Take further exploratory work studies.
  (ii) Summarize results of studies so far taken.
  (iii) Examine work study data for consistency.
  (iv) Take machine or process time studies (process operated or process controlled work measurement procedures only).
  (v) Critically examine any time study data for possible method improvement and implement any necessary changes (work measurement procedures only).

STAGE 7—SUMMARIZE INVESTIGATION AND DEVELOP REVISED PROCEDURES (work study staff).

(a) *Determine basic time values* (work measurement procedures only).

  (i) Classify elements into constant and variable duration.

  (ii) Take further time studies on the basis of this classification.

  (iii) Calculate basic elemental times of constant duration.

  (iv) Calculate basic elemental times of variable duration.

  (v) Determine relaxation allowances.

  (vi) Enter relaxation allowances against appropriate basic elemental times of constant duration.

  (vii) Enter relaxation allowances against appropriate basic elemental times of variable duration.

  (viii) Classify each work element according to the basis of its origin.

  (ix) Determine contingency allowance.

  (x) Add contingency allowance to all appropriate constant and variable elements and calculate standard times for these elements.

  (xi) Collect together all constant length elements to common bases of origin, determine the frequency of each to the common bases and total the results.

  (xii) Further collect together all work values classified under common bases to form the minimum number of elemental operations.

  (xiii) Determine the frequency of variation of elemental operations to the main basis of the time values.

  (xiv) Record the basic data in a convenient form.

(b) *Determine and check provisional time values* (work measurement procedures only).

(c) *Determine and reduce unoccupied time* (work measurement procedures on team work and operations involving machines or processes).

  (i) Analyse machine process times to establish a possible relationship between these and the product or process (process operated or process-controlled work measurement procedures only).

  (ii) Calculate unoccupied time.

  (iii) Reduce unoccupied time.

(d) *Calculate interference allowance* (process-controlled work measurement procedures only).

(*e*) *Determine and check final time values* (work measurement procedures only).

    (i) Time study the operation under theoretically optimum conditions (work measurement procedures on team work and process-controlled operations).

    (ii) Analyse provisional time values obtained above and adjust conditions if necessary (work measurement procedures on team work and process-controlled operations).

    (iii) Time study under further revised conditions if necessary (work measurement procedures on team work and process-controlled operations).

    (iv) Determine final time values (work measurement procedures).

    (v) Check final time values (work measurement procedures).

    (vi) Determine any special allowances (work measurement procedures).

Note For work measurement of operations of a unique, non-recurring nature, consider steps (iv) and (v) only, as explained in Chapter 7—page 86.

(*f*) *Develop improved methods of operation by critical examination of recorded data of existing methods and test on a trial basis. Simplify wherever possible.*

(*g*) *Compile work specifications of new methods.*

(*h*) *Obtain reference period.*

STAGE 8—DRAW UP DETAILED PLAN OF CHANGES AND CALCULATE NET FINANCIAL BENEFITS (work study staff).

(*a*) *Devise any necessary incentive bonus systems and design any special forms to be used for these.*

(*b*) *Design labour cost control system where this is not already in existence.*

(*c*) *Draw up plan of operation and conditions for installation of the new methods.*

(*d*) *Calculate achievable net cost reduction using reference period data.*

STAGE 9—REPORT ON PROPOSED CHANGES IN DETAIL (work study staff).

Summary of procedures

STAGE 10—SANCTION AND IMPLEMENT PROPOSED CHANGES (combined management action and work study staff).

(a) *Launch new method and give assistance at all phases to ensure successful installation within the original estimated time given in the survey report.*

(b) *Further assess revised methods as the application proceeds and improve and simplify wherever possible until the installation is complete.*

STAGE 11—REPORT ON FINANCIAL AND OTHER ACHIEVEMENTS (work study staff).

STAGE 12—MAINTAIN REVISED METHODS AND EXTEND IMPROVEMENTS WHEREVER POSSIBLE (combined work study staff and management action).

Part 3·—The procedure in detail

This part explains the procedural details given in Part 2 in more detail, and gives reference to other chapters where there are further explanations of some of the different techniques used. The adaptation of the procedure to suit varying conditions and applications is explained in Part 5 of this chapter.

STAGE 1—SELECT AREA FOR SURVEYAL

The initial selection of an area for surveyal is often motivated by top management due to some aspect of cost or apparent restriction of productive capacity over a period of time. Changing conditions may increase internal costs, affect output and reduce profit margins. Comparative examination of the company's accounts will often give a broad indication of where wastage is occurring, and production management may be aware of those areas which limit production. This does not mean that the work study department should never take the initiative. On the contrary, the maintenance of constant vigilance for sources of waste, and the subsequent reporting of all areas where there is scope for its reduction are amongst its prime duties.

Sometimes waste is quite independent of output. It may be happening in a maintenance department, tool room or office. A good work study practitioner should be able to assist in locating possible centres by carrying out an appraisal with the manager or overseer of the department. A general laxity of effort, or excess work-in-progress, is a fair indication of the existence of inadequate methods and low productivity—particularly if it is accompanied by persistent overtime working. Poor working conditions, ineffective maintenance of machinery and plant, or inefficiently operating plant, can reduce morale and often drastically depress productivity and lower quality standards.

These are all clear signs that there is a need for the application of work study.

STAGE 2—SURVEY AREA OF OPERATION

When an area has been selected as a possible source of waste, any other intention should first be discussed with supervision and the workers' representative before further action is taken (reference Chapter 1—page 5).

A survey of the activities is next carried out to discover more positively whether there is scope for increasing productivity and reducing costs by relieving obvious congestion, introducing improved methods, persuading operators to produce more by incentive, or a combination of these. At this stage, an assessment should be made of the present department productivity, the techniques which will have to be used (i.e. work measurement or method study—including examination of systems of documentation, etc.), and the time taken to carry out full investigation and bring about the necessary changes. Details of personnel, hours of work, gross wages earned and any other relevant details should also be obtained from authoritative sources.

If local or general congestion is apparent, it should not be automatically assumed that work study alone is the best approach. The main difficulty could be the lack of an efficient production control system. Work study combined with better control methods may then be the more effective. In fact, indiscriminate work study application without consideration of production balance may aggravate, instead of relieve some forms of congestion because, for example, by increasing the performance substantially of one department of an organization, one may flood succeeding departments with increases in work load which they cannot possibly accommodate.

STAGE 3—ASSESS POTENTIAL BENEFITS AGAINST IMPLEMENTATION COSTS

At this point, information collected during the survey stage, together with an assessment of the cost of implementing the proposals, is used to summarize the achievable benefits.

Consolidated results from the survey should give—

(a) Present assessed department performance or level of production with respect to the new method. This can be the actually observed level of the department performance, or an assessment of the scope of increasing performance levels by improved methods. For example, if better methods will reduce an operation time by one-quarter, then the performance should be correspondingly increasable by one-third.

(*b*) The present cost per unit of production (i.e. the cost per unit hour, per piece, per lb, per gallon, etc.) obtained by setting the present producing rates against the total internal production costs (labour cost, material cost, or overhead cost or a combination of all these).

Cost of implementing the proposals should be based on the following—

(*a*) The time required for the complete investigation, installation and maintenance of the new methods in terms of work study personnel, bonus clerks, etc.

(*b*) The cost of any capital equipment.

(*c*) The time and cost needed to install or re-install machinery and plant.

(*d*) The time required to train operators in any new methods or techniques.

(*e*) The amount of extra earnings expected to be paid to labour in return for higher productivity.

A conservative assessment should next be made of the future achievable levels of performance under the proposed operating conditions. This may take the form of assuming optimum performance levels being achieved by operators under a revised payment system and allowing for unavoidable losses in terms of scrap or lost time; or recording the performance of a proposed new type of machine or process.

The gross achievements of reduction in cost, increases in performance, reductions in scrap and other wastage, etc., are next calculated by comparing present with future unit costs.

The net achievement is finally obtained by setting the total implementation costs against the gross achievements. This may be considered in two stages—

(*a*) The time required to recover the full investment in implementation costs.

(*b*) The net benefits to be secured thereafter after consideration of the cost of maintaining the new methods.

It is on these two factors alone that the final recommendations are made.

STAGE 4—REPORT ON RECOMMENDATIONS

Recommendations to management on the economic viability of further investigation and action, should now be summarized in the form of a report. Although dependent on present conditions, a proposition can usually be considered to be worth while if the net gains will fully recover the investment costs within two years or less. (By net gains is meant gross benefit less maintenance cost.)

Reports should be informative and authoritative but extremely brief. The body of the report should be confined to one, or at the most two, single sheets of A4 size paper, reference being made to appendixes for any necessary support data, sources of information and any necessary calculations. It is important to state quite clearly whether or not recommendation is being made to proceed with the investigation (reference Chapter 17).

STAGE 5—SANCTION INVESTIGATION

Since there is no more effective way of stifling enthusiasm than indecisive management, it is important that reports submitted are acted upon by management within a short time of receiving them, the only excusable delay being the independent checking of calculations and sources of information. If substantial net cost reduction can be secured within a reasonable period of time—particularly if little capital expenditure is involved—the project should be sanctioned without hesitation, assuming that the staff is free to proceed with it without delay.

STAGE 6—CARRY OUT DETAILED INVESTIGATION

This is the stage during which detailed information of existing working methods is collected and recorded. No more detail should be set down than is necessary for the final objective, and economy of investigation time must be carefully preserved throughout to ensure that the time taken is within the limits allowed for in the survey estimate. To prevent duplication of effort and wasted time, investigation should be conducted as three separate operations, the information collected in each of them being used to plan for the next.

(a) *Further investigate prior to work study*
This is the first investigation operation.

(i) Discuss the investigation with supervision and the workers' representative.

As human considerations rank high in work study, it is essential that this discussion is carried out before any work is done. The procedure for so doing is fully explained in Chapter 1.

301

(ii) Select operators for preliminary study.

Much time can be saved if the more competent of the operators are selected for preliminary observation. Supervision can greatly assist in this choice (reference Chaper 3).

(iii) Consult technical staff if the equipment or process is not fully understood.

On many industrial operations there may be technical processes which are unfamiliar to the observer. Also, there may be limiting time cycles which are necessary to complete the processes to satisfactory quality standards—such as cure times on rubber and plastic moulding, immersion times on metal plating, cooling cycles on metal casting, feed and speed times on metal cutting, etc. In these cases, competent technical staff should be consulted until the significance of the machines or processes, and the effect of altering any of the cycle times is fully understood.

(iv) Draw up preliminary work study programme.

At this point, a fair appreciation of the personnel and the nature of the greatest bulk of the processes and products handled in the department should have been obtained. This will enable broad plans to be drawn up to decide which operators and processes should first be subjected to investigation, in order to obtain a deeper understanding of the department's activities, and allow concentration in those areas which are most important to its function.

It may have been noticed that due to local congestion, the layout of the department or work place can be improved simply. The programme would then include broad method study of the areas of greatest congestion; the products and operators selected for preliminary investigation; and machines or processes apparently malfunctioning which are earmarked for more concentrated time or method study investigation. If mainly work measurement is involved, much of the time study can still be programmed to take place even before any revised methods have been implemented. For in many cases, broad method study will only tend to alter the incidental work, and the bulk of the elements which constitute the main operation may remain unaffected.

(b) Carry out initial work study observations

This is the execution of the preliminary investigation programme drawn up in the previous step and the consolidation of work done so far to form a final investigation programme.

(i) Draw site plan of workplace layout.

This is an essential preliminary to any investigation and will form a useful record for reference or as a basis for method study. In many cases a rough sketch may be adequate if there is insufficient time or facilities to prepare a scale plan (reference Chapter 12).

(ii) Take exploratory observations of present working methods.

This is the systematic collection of information from various sources to gather a more comprehensive idea of the operating conditions. It could involve determining and recording the reason for operations and the order and path of movement of materials, paper work, etc., confining these to broad terms at this stage (reference Chapter 12). It could also involve the recording of machine speeds, or operating cycles on any special processes. Direct observation of these should always be used in spite of helpful evidence which may be supplied in good faith by other authorities. Spindle speeds should be checked with a revolution counter and stop watch, the revolutions per minute being calculated afterwards. Operating speeds of machine slides should also be checked similarly over a measured length. Where very high speeds exist, it may be necessary to use a stroboscopic disc or stroboflash, particularly when the extra load imposed by a revolution counter will slow the machine. Tachometers are not generally recommended because of the difficulty of obtaining true average speeds. The details of such experiments should be recorded on time study observation forms (reference Chapter 10).

If work measurement is being carried out, a few time studies of very short duration would be taken on selected operators (reference Chapter 3).

(iii) Report on any preliminary broad method changes needing management action and ensure their prompt installation. This step covers the construction of outline process charts and flow diagrams and the preparation of proposed rearrangement of workplace and workshop layout of a simple nature. Plans involving major plant movements should not be considered at this stage. Discuss any proposed changes with shop supervision before finalizing the plans.

Correct settings of machine speeds, process operating cycles, etc. should be established after discussion with technical staff, by actual experiment, or consultation of machine makers' handbooks, etc. Proposals for any changes of this nature, or of any other broad method modifications should be briefly reported to management to ensure their implementation at the earliest possible opportunity (reference Chapter 12).

(iv) Plan main work study programme.

Sufficient information should have been collected by this time to enable the whole programme to be planned in line with the limits set out in the original survey. The programme should be arranged around the known main products and processes in the department. Where micromotion study is to be carried out, arrangements should be made to use ciné and other cameras, cyclegraphic apparatus, etc.

(c) Carry out main work study observations

Although during this investigation period most of the detailed information is collected from the activity centre, a certain amount of summarizing must also be done at the same time to avoid unnecessary duplication.

(i) Take further exploratory work studies.

More detailed studies are taken of particular areas of activity. Data is collected directly by observation and questioning, using a time piece, activity sampling or other methods if work measurement is involved. Collection for micro-motion study is carried out by ciné camera, or by camera and cyclegraph apparatus. If carrying out work measurement, time studies should be of a longer duration than those taken initially (reference Chapters 3 and 8 for work measurement collection techniques, and Chapter 12 for method study collection techniques).

(ii) Summarize results of studies so far taken.

Time studies should be summarized at the earliest opportunity after taking them—using the procedures described in Chapter 3. No data is entered on summary sheets at this stage.

If *method study* techniques are being used, the data so far taken is summarized in the form of process charts, flow diagrams, string diagrams, simo charts, etc.

(iii) Examine work study data for consistency.

Time study data should be constantly examined to check that all work elements end on the same break point. At least one study of long duration should be taken in addition to studies covering starting and finishing periods. This is to ensure collection of all incidental work and provide reference for checking of time values (reference Chapter 3).

Method study charts must also be checked against the originating sources of information to ensure absolute completeness. (reference Chapter 12).

(iv) Take machine or process time studies.

This stage is only necessary when carrying out work measurement of those process-operated or process-controlled operations where the cycle times tend to vary with different products. Such conditions occur on lathes, milling machines, etc. where varying diameters, lengths, depths of cut, type of material, class of finish, etc. may vary the cutting times. Other examples are on textile machines, where the size of yarn, weight of material wound, etc. will influence the cycle time, or on die-casting, plastic injection moulding, etc. where the shape and size of the product will affect the injection and cooling cycles.

The operation consists of recording net machine or operation cycle times together with characteristics of the product or material (reference Chapter 10).

(v) Critically examine any time study data for possible method improvement and implement any necessary changes. Whenever work measurement is carried out, the data collected must be constantly appraised to discover any improvements that can be simply applied. Sometimes method improvements can be discovered by comparison of time study data which could escape the notice of an observer carrying out pure method study without time values (reference Chapter 4).

STAGE 7—SUMMARIZE INVESTIGATIONS AND DEVELOP REVISED PROCEDURE

This is the stage during which investigations are summarized in a convenient form, subjected to analysis and developed into a revised procedure which, when put into practice, will achieve the benefits claimed and reported during stage 4.

When conducting work measurement of a unique, non-recurring operation, the simplified procedure explained in Chapter 7 (page 86) is used in place of steps (*a*) to (*e*).

(*a*) *Determine basic time values*

This is a period of consolidation and crystallization of the work measurement programme leading to the determination of basic data which can be used to compute final time values.

(i) Classify elements into constant and variable duration. This consists of carefully scrutinizing elemental times on all time studies so far taken, noting those which tend to remain constant irrespective of variations in the product or process, and entering these to *study summary—constant elements—*

sheets (Fig 17) and those which tend to vary to *study summary* and *register of variable elements sheets* (Fig 18).

Note—*This is the first entry of any data from time studies to summary sheets in the whole procedure* (reference Chapter 4).

(ii) Take further time studies on the basis of this classification.

Examine the entries on the *summary sheets* critically and concentrate further time study only on those elements not yet observed, and those elements on which insufficient observations have been taken—being all those which are deemed necessary to compute final time values. This policy must be pursued vigorously, continually reviewing the situation to ensure that no more observations are taken than is absolutely necessary (reference Chapter 4).

(iii) Calculate basic elemental times of constant duration. This consists of selecting the constant duration elemental times previously entered on the *summary sheets* (reference Chapter 4).

(iv) Calculate basic elemental times of variable duration. Carefully examine all variable duration elements entered to the summary sheets and determine the possible basis of variation of each to some easily identifiable feature of the product, machine, or process. Examples are length, width, depth, area, weight, variety of shape, type of machine or machine setting, machine speed, type of material, degree of finish, type of tool or jig used, etc. Graph or list the variable elements according to the bases of origin and select variable element times *direct from the graphs or other classifying media* (reference Chapter 4).

(v) Determine relaxation allowances.

These are computed as described in Chapter 5, classifying them into groups to be used over a range of elements to save needless computation.

(vi) Enter relaxation allowances against appropriate basic elemental times of constant duration. This is explained fully in Chapter 6.

(vii) Enter relaxation allowances against appropriate basic elemental times of variable duration. This is a similar procedure to that above and is again fully described in Chapter 6.

(viii) Classify each work element according to the basis of its origin.

All work elements have a reason for being performed. An element "move tray" exists because there are trays which need to be moved; "paint discs" exists because there are discs which need to be painted, etc. Both constant and variable duration elements should be classified in this way (per tray, per disc, etc.) (reference Chapter 6).

(ix) Determine contingency allowance.

This is obtained by extending all elements "per day" or

"per shift" with the addition of relaxation allowances and adding these together. This is then expressed as a percentage of the working day or shift (reference Chapter 6).

(x) Add contingency allowance to all appropriate constant and variable elements and calculate standard times for these elements.

At this point the contingency allowance is added to all elements (except the contingency elements themselves) and standard times for each obtained by extension of both the relaxation and contingency allowances to the basic elemental times (reference Chapter 6).

(xi) Collect together all constant elements to common bases of origin, determine the frequency of each to the common basis and total the results. This consists of collecting together all elements by common basis (e.g. per disc, per tray, per trolley, etc.) determining the frequency of each to these bases (e.g. once per trolley, twice in five trays, etc.), multiplying each element by the frequencies expressed as a vulgar fraction and adding the extended values together. The procedure is explained in detail in Chapter 6.

(xii) Further collect together all time values classified under common bases to form the minimum number of elemental operations.

This consists of a further rationalization of elements to form time values on a minimum number of bases. If, for example, there is a time value per trolley and a time value per tray, and it is noticed that there are always the same number of trays loaded to a trolley, then the two time values can be combined as one time value "per tray".

The purpose of this is to reduce the final data to minimum proportions whilst still preserving the original accuracy (reference Chapter 6).

(xiii) Determine the frequency of variation of elemental operations to the main basis of the time values. This is the determination of the frequency of happening of time values with respect to the common bases (e.g. the frequency which the work values per tray, per charge of spray gun, etc. occur to the actual pieces produced, i.e. discs). Frequencies may be fixed or variable according to the type or size of the final pieces produced (reference Chapter 6).

(xiv) Record the basic data in a convenient form. It is important to have all the basic data filed for quick reference in such a way that it can be easily understood for computation of time values (as well as being traceable to its originating source) (reference part 4 of this chapter).

(b) *Determine and check provisional time values*
At this stage, it will be necessary to compute certain time values from the basic data so far compiled and check them against the original time study to ensure that there have been no arithmetical or mathematical errors (reference Chapter 7).

The work values will not be in their final form until consideration has been given to any possible modifications in the way the work is carried out, due to the incidence of unoccupied time, machine interference, or other factors introduced in an attempt to reduce the work content or simplify the procedures.

(c) *Determine and reduce unoccupied time*
Unoccupied time is caused by operators being forced to wait for a process to be completed which is longer in time than the operation which they themselves are doing when they are working at standard performance.

This part of the procedure consists in determining this time and so rearranging the work as to reduce it to a minimum. It is considered in three steps as follows—

(i) Analyse machine process times to establish a possible relationship between these and the product or process. This will only be necessary where machines and processes are used which tend to be controlled by the operation because of automatic or semi-automatic cycle times. It is similar to the procedure carried out when analysing elemental times of variable duration. Data obtained by time study during stage 6 (c) (iv) is subjected to careful analysis to determine a firm relationship between these and the product or process. Examples of possible bases for variation are: material used; lengths, type and size of yarn or thread used (in textiles); size, shape, weight, etc. of parts produced (die casting, plastic injection moulding, lathe turning, etc.) (reference Chapter 10).

(ii) Calculate unoccupied time.
Unoccupied time is obtained by comparison of a regulating process time with other operating standard times. It can most conveniently be shown by constructing a *multiple activity chart*. (reference Chapter 9). On machine or process-controlled operations it will first be necessary to segregate time values into an amount which has to be performed outside the process cycle and that which can be performed within the cycle. The unoccupied time is then the difference between the inside work and process time (reference Chapter 10).

(iii) Reduce unoccupied time.

This can be reduced either by shortening the controlling cycle or by providing other work to be done in the unoccupied period. Where team work is involved, it can be achieved by reallocation of the work to shorten the governing operation (reference Chapter 9).

On machine or process controlled operations other work can sometimes be provided by allocating more than one machine to an operator. This may, however, raise problems of machine interference. This will have to be considered separately in the next step.

If different processes or products alter the process times, these should be computed from the data prepared during step (i) above (reference Chapter 10).

(*d*) *Calculate interference allowance*

Also known as "synchronization", this allowance is needed to cover the incidence of loss which occurs during the operation of a number of automatic or semi-automatic machines or processes due to one or more of them ceasing to operate (i.e. completing their cycles) before others, needing attention, can be restarted.

The procedure should be as follows—

(i) Determine the maximum number of machines or processes which can be allocated to an operator by dividing the total standard time per cycle into the machine cycle time less the inside work per cycle, and rounding off to the lower whole number.

(ii) Determine the theoretical interference losses by using the formulae given in Chapter 10 at minimum unoccupied time conditions, and at various other conditions of increased unoccupied time by reduction in allocation of the number of machines.

(iii) Obtain theoretically optimum operating conditions of the machines by balancing the maximum tolerable unoccupied time with the minimum tolerable interference loss on an operating cost basis.

Theoretically calculated conditions are then checked against practical achievements in the next step (reference Chapter 10 for detailed procedure and formulae used).

(*e*) *Determine and check final time values*

This is the process during which final work values are determined after considering further cost reduction by possible rearrangement of the work pattern, especially where there has been incidents of unoccupied time and machine interference.

(i) Time study the operation under theoretically optimum conditions.
This step applies to work measurement applications where rearrangement of duties has been proposed in an attempt to reduce the work content of the operation, and covers time study observation of the new conditions specifically arranged for the purpose in order to test the effect of the changes on a practical basis (reference Chapter 7, and Chapters 9 and 10 for team work and process-controlled operations).

(ii) Analyse provisional time values obtained above and adjust conditions if necessary.
This is the detailed analysis of the results of the time study taken above and the further rearrangement of the work to reach optimum conditions if this is necessary (reference Chapter 7 and Chapters 9 and 10 for team work and process-controlled operations).

(iii) Time study under further revised conditions if necessary. If any but small adjustments have been made in step (ii) above, it may be necessary to carry out more time study under further revised conditions (reference Chapters 9 and 10).

(iv) Determine final time values.
This is the period during which selected data is used to compile final time values.
Where there have been no major changes in the method of operation brought about by the incidence of unoccupied time, machine interference or other rearrangement of duties to reduce the work content, this will have been carried out in step (*b*) above (reference Chapters 7, 9 and 10).

(v) Check final time values.
Final time values should be checked against original time studies or special production studies if this has not already been done in steps (*b*), (*e*) (i), (ii) and (iii) above. Once basic data and final time values have been checked and found correct by this means, no more checks of this nature should be necessary in future (reference Chapters 7, 9 and 10).

(vi) Determine any special allowances.
Sometimes, due to varying conditions of working, it may be necessary to obtain certain other work values in the form of special allowances in order that all work done by operators is properly measured. Examples of work in this category are excess work allowances due to temporary faults in working methods and special allowances for loading a machine with

310

work prior to its full operation (reference Chapter 10 for special starting and other allowances on machine-controlled operations).

(*f*) *Develop improved methods of operation by critical examination of recorded data of existing methods and test on a trial basis. Simplify wherever possible*

For *method study* applications, this is the critical examination of process charts, flow diagrams, string diagrams, simo charts, documentation methods, etc. and the reduction of these to the simplest possible terms consistent with their effective operation. This is followed by the experimental testing of these in practice on a trial basis (reference Chapter 12).

For *work measurement* applications, this is the appraisal of synthetic data and time values with a view to reducing their number, bearing in mind that the greater the number of time values used by operators in any one week, the wider the limits of "rounding off" of these is possible (whilst still retaining sufficient accuracy for bonus incentive payment). This is particularly true if the incentive is so designed to even out, or eliminate, fluctuations in pay when there are relatively small fluctuations in effort. This is a most important consideration, since it will reduce the maintenance costs of an incentive bonus system to minimum limits and tend to eradicate one of the major causes of unrest, whilst still retaining the incentive to produce at optimum effort (reference Chapter 11—"accuracy of standard times").

(*g*) *Compile work specifications of new methods*

This is the qualification of revised methods and time values by a full and detailed description of the operations involved based on the new conditions (reference Chapter 11).

(*h*) *Obtain reference period*

Necessary reference period data concerning conditions before the investigation began—in the form of gross wages, hours of work, material costs, overhead costs, output details, etc.—should have been collected from reliable sources at this point. If no proper records had existed before beginning the project, temporary ones should have been set up at an earlier stage.

The data is then used to compile reference period costs and conditions for comparison purposes (reference Chapter 17).

STAGE 8—DRAW UP DETAILED PLAN OF CHANGES AND CALCU-LATE NET FINANCIAL BENEFITS

This is the stage when the method of implementing the final changes is drawn up into a procedure for their ultimate installation.

(*a*) *Devise any necessary incentive bonus systems and design any special forms to be used for these*

It is important that incentive bonus systems are designed in such a way that it persuades operators to direct their efforts towards the objectives. If quality of the product or material waste control is important, the incentive should be constructed around these factors as well as high productivity.

Fluctuations in pay for small fluctuations in effort should be avoided. Carefully designed bonus incentive systems on this basis combined with the minimum number of time values achieved during Stage 7 (*f*) can greatly simplify conditions. This will reduce maintenance costs and ensure that the system can be easily understood by labour.

Work sheets for recording output, lost time, etc. must be designed to minimize the clerical operations (reference Chapter 14).

(*b*) *Design labour cost control system where this is not in operation*

Labour cost control can only be operated successfully where some form of work measurement is in existence or is about to be installed. Since it is vital for the maintenance of cost reduction and a valuable indicator to the source of further waste it should be implemented at the earliest opportunity and certainly no later than this stage in a work measurement application.

The forms used should be so designed that the absolute minimum of information is presented and issued only to those authorities in the management team who are in a position to act on any faults revealed by it (reference Chapter 15).

(*c*) *Draw up plan of operation and conditions for installation of new methods*

In drawing up this plan consideration may include the following—

(i) Installation or re-installation of machinery and plant to bring about improvements in workshop or workplace layout (reference Chapter 12).

(ii) Improvements to mechanical handling methods (reference Chapter 12).

(iii) Installation of improved storage facilities (reference Chapter 12).

(iv) Programmes for operator training in new methods of work performance (reference Chapter 12).

(v) Increases in pay which may be necessary for operators to persuade them to achieve the target increases in productivity (reference Chapter 17).

(vi) Time required and conditions to allow operators to learn any new methods and achieve optimum productivity (reference Chapter 14).

(vii) Proposed conditions for the operation of any bonus incentive system for direct and indirect labour (reference Chapter 14).

(viii) Estimated time required for full installation (this should be in line with the recommendations made during Stage 4).

(ix) Design of any special forms to be used for rapid computation of time values from synthetic data (reference Chapter 11).

(*d*) *Calculate achievable net cost reduction using reference period data*

This is the process of setting the gross savings achievable against the total installation and maintenance cost to obtain the net benefits.

Consideration must be given to costs of—

(i) Capital equipment.

(ii) Installation or re-installation of any equipment.

(iii) Work study personnel assigned to the work.

(iv) Possible increases in wages necessary to secure optimum conditions.

(v) Final maintenance.

These are then set against the gross savings achievable to give the final result (reference Chapter 17).

STAGE 9—REPORT ON PROPOSED CHANGES IN DETAIL

This is the arrangement of the information collected in the previous stage in the form of a report to management setting out the proposals and the benefits to be secured. The report should be brief and to the point, possible financial achievements clearly set out, any supporting data and calculations being attached as an appendix with reference to all the sources of information (reference Chapter 17).

STAGE 10—SANCTION AND IMPLEMENT PROPOSED CHANGES

This stage needs combined action by management and work study staff, since not only will the scheme need to be approved in full before installation, but it will need to be supported by management throughout.

(*a*) *Launch new method and give assistance at all phases to ensure successful installation within the original estimated time given in the survey report.*

Management sanction should be given without hesitation to proposals submitted at this stage providing there has been no significant change from the orginal proposals made in the survey report.

313

Before any changes are made, a discussion should be held with both supervision and the workers' representatives to explain the operation of the new methods. It is desirable that such meetings are chaired by a representative of management to give them full authority.

As the scheme progresses—particularly where incentive bonus payments are involved—work study staff should go to great lengths to explain to labour how to achieve the targets set, and to help them in every possible way. It would be worth while in many cases spending a whole day with them to give encouragement and assistance. Great patience and tact must be exercised at this period to bring the scheme to fruition, no effort being spared to attempt to achieve optimum results within the estimated time originally set in the original survey report (reference Chapters 14 and 12 concerning work measurement and method study applications respectively).

(*b*) *Further assess revised methods as the application proceeds and improve and simplify wherever possible until the installation is complete*

As the application of new methods continues, improvements will very often suggest themselves and it may be possible to simplify some of the procedures. These modifications should be incorporated until the best possible methods are in operation.

STAGE 11—REPORT ON FINANCIAL AND OTHER ACHIEVEMENTS
Financial and other achievements should be calculated weekly and brief reports on the progress of the application regularly sent to management, culminating with a final report on completion of the project (reference Chapter 17).

STAGE 12—MAINTAIN REVISED METHODS AND EXTEND IMPROVEMENTS WHEREVER POSSIBLE
It is obviously most important that the revised methods are properly maintained within the originally estimated maintenance cost. Work measurement applications should never be allowed to deteriorate into time study of all new operations followed by individual rate setting, but always computed from existing synthetic data, together with any newly studied data which has been incorporated.

Improvements in method must always be initiated by the work study department. This is most important, since any measured work values must be aligned with method improvements immediately they take place.

Where there are known improvements that can be made to methods as a second stage, this stage may in fact be the commencement of another project starting at stage 1.

Part 4—Referencing procedure

It is most important that there is an efficient filing system for work study data so that it can be referred to quickly—particularly when carrying out work measurement, and original time studies are required to be traced from final time values or synthetic data. To achieve this, it will be necessary to print a series of standard forms from a printing block or stencil, and operate the data referencing procedures described below.

FINE WORK MEASUREMENT TO PRODUCE SYNTHETIC DATA

This procedure is recommended when carrying out work measurement by time study with a stop watch, and the work is being divided into time intervals of approximately ten to fifty centiminutes (six to thirty seconds).

The filing system is in 5 groups.

(a) *Time study*

Data is recorded either on *work study observation and record sheets* (Fig 10), if it is repetitive, and *work study observation sheets* (Fig 9), if it is not. Every sheet is headed with a study number allocated from a *study register* (Fig 11).

The data recorded during study is summarized on *work study observation and record sheets* and each element is given a reference letter. Resulting data are finally entered to a *work study top sheet* (Fig 13), the same reference letters being used for each element entered. The whole is then stapled together in the following order, all sheets, except the top sheets, being numbered sequentially. Time studies should then be kept filed in numerical order after entry and extraction of data (reference Chapter 3).

Work study top sheets
Work study observation and record sheets
Work study observation sheets

(b) *Summary data*

Elements of constant duration are entered from *work study top sheets* to *study summary constant elements sheets* (Fig 17), all of which are numbered with a set-up number, the sequential order of sheets being indicated by a letter reference at the head of each. Sheets are designed to have pre-printed summary reference numbers 1 to 8 on each, so that elements entered can be specifically referred to by set-up, reference letter, and summary reference number. For example, an element of constant duration from set-up No. 10, *summary sheet* B, summary reference No. 6, is referred to as element 10B6, and so on.

Elements of variable duration are entered from *work study top sheets* to *study summary and register of variable elements sheets* (Fig 18), all of which are numbered with a set-up number, the sequential order being indicated by a number reference against the prefix letter *V* printed at the head of each.

Since sheets have pre-printed reference numbers 1 to 25 on each, any elements entered can be referred to specifically by set-up, sheet number, and reference number. For example, an element of variable duration from set-up No. 10, *summary sheet* V3, reference number 7 is referred to as 10/V3/7 and so on.

All elements on either type of summary sheet are clearly referenced back to their originating time study numbers at the time of entry.

Variable elements may be selected by comparison on *graph sheets* (Figs 19 and 24) both before and after adding relaxation and contingency allowances. All these sheets are headed with the set-up number and sequential number against the prefix letter *G* printed at the head of each. Values selected from these graphs and entered to *study summary and register of variable elements sheets* are referenced back to the originating graph sheets.

Selected elemental basic times of constant duration are added from *study summary-constant elements sheets* to the *register of constant elements sheets* (Fig 21), entering against each the summary sheet reference number (10B6, etc.) as explained above. Each *register sheet* is headed with the set-up number and sequentially numbered. Since it is also pre-printed with reference numbers 1 to 25, elements entered to it can be referred to by set-up, sheet number and reference number, e.g. 10/1/12 means element number 12 on sheet 1 on set-up number 10, etc. These numbers are entered to the *study summary —constant elements sheets* to provide forward reference.

When elements are further grouped together into common bases of origin in standard time form, they are entered to *collection of elements sheets* (Figs 22 and 23) together with their register reference numbers (constant or variable). *Collection of elements sheets* are all headed with a set-up number and a sheet number against the prefix letter *C*.

Final computation and recording of basic data is carried out on *data sheets* (Figs 25 and 33), elemental operations from *collection of elements sheets* being referred back to their originating sheets. For example, if it was taken from sheet C1 of set-up number 10, this would be entered as 10C1, etc. Further collection of basic data to form final work values for cross-checking by time studies is also entered to *data sheets* (Figs 25 and 33).

Site plans, workplace layouts and other sketches, and any necessary calculations for checking work values against original time studies or special production studies are also entered to *data sheets* (Fig 25), these being headed with the appropriate set-up and sequential number. Where site plans are large and bulky, it is usually more convenient to file these separately, although reference must be made to the drawing numbers of these on *data sheets* which form part of the summary data. Finally, a contents sheet is prepared using a *data sheet* on which is set out a complete list of all the sheets used together with their reference numbers.

The whole is then stapled together in the following order, the contents sheet forming a top sheet to the summary—

Contents on *data sheets*
Cross-checks from time studies on *data sheets*
Final work value computations used for cross-checking on *data sheets*
Basic and synthetic data on *data sheets*
Collection of elements sheets
Register of constant elements sheets
Study summary and register of variable elements sheets
Graph sheets
Study summary—constant elements sheets
Site plans, workplace layouts on *data sheets*

When carrying out work measurement of machine or process controlled operations, data from time studies of the machine operating cycles and speeds, etc. are entered to *study summary and register of variable elements sheets* and compared on *graph sheets*, results of any formulae being computed and recorded on *data sheets*. *Data sheets* are also used to show calculations involved in arriving at interference allowances.

Work measurement of team work and process controlled operations which involve the construction of multiple activity charts can be drawn on either *graph sheets* or *data sheets*.

Where charts and graphs cannot be drawn using standard sized sheets, larger ones can be used and folded map-wise to the required size.

If this referencing system is used, it will be possible to trace any data at any point in the proceedings back to its originating source— a most valuable asset when checking accuracy, or when the validity of work values is questioned by labour.

(*c*) *Relaxation allowances*
These are computed on *relaxation allowance computation forms* (Fig 20), which are then filed in numerical sequence in a separate file.

317

Summary of procedures

(d) Project records

These are records of work specifications, details of synthetic data, regulations for the operation of incentive bonus systems and any other records to which supervision and labour should have free access. *Data sheets* can be used exclusively for these, and kept filed in departmental and chronological order, reference being made to set-up numbers where appropriate.

(e) Reports

Since more than one set-up may be involved in a work study project, it is advantageous to allocate each a project number. Reports and other correspondence can then be filed in chronological sequence under this number.

FINE WORK MEASUREMENT OF NON-RECURRING WORK

This simplified procedure is used for non-recurring work of a unique nature. It only differs from the previous procedure in the method of summarizing data, all other steps remaining the same.

Summary Data

Since all elements are of fixed duration, they are entered from the *work study top sheets* directly to *study summary-constant elements sheets* (Fig 17), and after selection to a *register of constant elements sheet* (Fig 21), relaxation allowances being computed as before on a *relaxation allowance computation form* (Fig 20), and added to the elements, the final work values being computed on the *register of constant elements sheet*.

A *data sheet* (Figs 30 and 31) is then used as a contents sheet and all sheets stapled together in the following order—

Contents on *data sheet*.
Work value computation on *register of constant elements sheet*.
Study summary—constant elements sheets.
Site plan or workplace layout.

BROAD WORK MEASUREMENT

This procedure is designed for use when carrying out work measurement either by time study using a wrist watch or wall clock; direct observation without direct timing; analytical and comparative estimating; activity sampling; estimates from examination of records; or by a combination of all these.

(a) Data collection

Where data is being collected by time study using a wrist watch or similar device, the referencing procedures and forms are exactly the same as those used for fine work measurement.

318

Collection of data by direct observation without timing is very similar—except that the individual times for each phase or work element have to be estimated, the total estimates being made equal to the elapsed time less any lost time. From then on, the procedure is the same as above.

Activity sampling is carried out on a *work study observation and record sheet* (Fig 30), the results being summarized and entered to a *work study top sheet* (Fig 28).

Analysis from records is carried out on *work study observation and record sheet* and the final results entered to *work study top sheets*, the records themselves being stapled to the study sheet and forming part of it.

(b) Summary data

Where data can be collected directly using time study with a wrist watch or wall clock, the procedure for summation of data and the forms used for so doing are exactly the same as that used for fine work measurement, the only difference being that the work elements will be considerably longer in duration.

Where this method is not practical for any reason, and a combination technique is adopted by direct observation without timing, analytical and comparative estimating, analysis of records, and activity sampling, the procedure will have to be slightly modified.

Analytical estimating is carried out on *register of constant elements sheets* (Fig 27), elemental times of the estimates being entered under the column headed "standard minutes per 100 occs" and the total estimated time entered at the foot of the form. Each of these is then headed with the set-up number, the sheets being also numbered in sequence.

Direct observation without timing will probably consist of work elements inclusive of rest allowance. Constant and variable duration elements of this character should be entered under the column "standard minutes per 100 occs" on *study summary and register of variable elements sheets*. Constant duration elements can be sequentially numbered V1 onwards and variable duration elements V101 onwards to separate the two groups.

Estimated times for complete jobs and other details for activity sampling studies and analysis of records, are entered directly to *data sheets*, reference being made to the originating study numbers on those sheets.

Selected elemental times of constant duration originally obtained from time studies by wrist watch or wall clock (where these have been taken), are entered from *study summary constant elements sheets* to *register of constant elements sheets* and relaxation allowances added. Contingency allowances are not normally considered when using broad work measurement.

Elemental times of variable duration entered to *study summary and register of variable elements sheets* are also converted to standard times by the addition of relaxation allowance.

Comparison of standard elemental times of constant duration from both time studies and analytical estimates is made by entering these to *study summary-constant elements sheets*, finally selected elemental standard time being then entered to a fresh set of *register of constant elements sheets*.

Comparison of variable elements obtained from various sources can also be made on a common basis (i.e. all with relaxation allowance added) using *graph sheets* or *data sheets* and referencing the data in a similar manner to that described under fine work measurement, finally selected elemental times being re-entered to the *study summary and register of variable elements sheets* under the column headed "Standard minutes per 100 occs.".

Time values are checked against overall times obtained from activity sampling studies, studies by direct observation throughout the timing, and studies by analysis of records, using *data sheets* for the purpose.

Synthetic data and final work values are recorded on *data sheets*, and a contents sheet prepared on the same type of sheet, as described under fine work measurement. The whole set-up is then stapled together in the following order—

Contents on *data sheet*.

Cross checks from activity sampling studies, analysis of record studies, general observation studies and time studies on *data sheets*.

Final work value computations used for cross checking purposes on *data sheets*.

Basic synthetic data on *data sheets*.

Collection of elements sheets

Final register of constant elements on *register of constant elements sheets*.

Study summary and register of variable elements sheets.

Graph sheets.

Summary of constant elemental standard times on *study summary —constant elements sheets*.

Initial register of fixed elemental standard times on *register of constant elements sheets*.

Summary of basic elemental times on *study summary—constant elements sheets*.

Site plan or workplace layout on *data sheets*.

Relaxation allowances, project records and reports are filed in exactly the same way as for fine work measurement.

FINE AND BROAD METHOD STUDY

This procedure is recommended for all method study except micro-motion study.

(a) Collection of data

Data is recorded on *work study observation sheets* (Fig 9) and analysed results entered either to a *work study top sheet* (Fig 13) or a *process chart sheet* (Fig 46). Each study is allocated a study number from the *study register* in a manner similar to that in which fine work measurement is carried out. The completed study is then stapled together in the following order—

> *Work study top sheet* or *process chart sheet*
> *Work study observation sheets*

All sheets except *top sheets*, are numbered sequentially and all data recorded referenced by being given a letter reference, these being duplicated on the *top sheets*—the *process chart sheet* being regarded as a *top sheet* when this is used.

(b) Summary data

Process charts, flow charts and site plans constructed from the collected data are drawn on *data sheets* or on large sheets folded map-wise to the same size. The whole is then stapled together with a contents sheet on top and filed under set-up numbers allocated from the *set-up register* (Fig 15). Reference is carefully made to the originating study numbers on all data.

Project records (which include full work specifications) and reports are filed in chronological order under a project number in the same way as when carrying out work measurement.

MICRO-MOTION STUDY

(a) Collection data

Data for *simo charts* is recorded on ciné film, and cyclegraphs or chronocyclegraphs are recorded on still photographs. These should be allocated a set-up number from the *set-up register* (Fig 15) and filed under this reference.

(b) Summary data

Simo charts and other analytical data are constructed on data sheets of such a size that they can be folded map-wise to the same size as a standard *data sheet*.

Summary of procedures

Completed data should be filed under set-up numbers in the following order—

Contents on *data sheet.*
Simo charts.
Analysis data on *data sheets.*
Cyclegraphs or chronocyclegraphs.

Project records and reports are filed under department and project number in the same way as that used for work measurement.

Part 5—Adaptations of the procedure to suit different applications and conditions of work

The procedures explained in parts 2 to 4 of this chapter are intended to be fully comprehensive for the application of work study to cost reduction. After discovering possible sources of waste, and carrying out an economic appraisal during Stages 1 to 4, however, it will be necessary to adapt the procedure to suit the particular conditions. This consists of selecting those parts which apply to the techniques decided upon.

This part contains a number of these adaptations which can be used to cover most the different situations likely to be met in practice.

Case 1—Work measurement of repetitive or semi-repetitive operations performed with the minimum assistance of powered machinery or plant

This procedure is suitable for measuring operations—using fine work measurement—which are either repetitive or semi-repetitive in nature, but which are not performed with the aid of powered machinery and on which there are some common bases for comparison one with the other. The procedure is also intended to cover cost reduction by combined work measurement and bonus incentive.

Typical examples include the many light assembly operations which are frequently met in those industries using batch or mass production, except those which are performed by a team of operators.

STAGES 1–5
During the survey stage, a close assessment would have to be made of department productivity, and the potential cost reduction then estimated, assuming the achievement of optimum departmental performance in the future. This would have to be set against any necessary pay increases to provide sufficient incentive to the operators (reference Chapter 17).

STAGE 6

This stage would consist mainly of time study and would follow the full procedure as set out in parts 2 and 3 except that steps (*a*) (iii) and (*c*) (iv) are not necessary.

STAGE 7

This stage would consist of all those activities as set out in the full procedure, but omitting (*c*) and (*d*).

STAGE 8

This stage would have to be carried out in full, including preparation for the installation of labour cost control.

STAGES 9–12

No effort should be spared during stage 10 to enlighten and assist operators in how to achieve target earnings by attaining optimum productivity. It is most important that the system is efficiently maintained by continual use of synthetic data and not allowed to deteriorate into rate setting by individual time studies.

REFERENCING PROCEDURE

The referencing procedure to be used is as that described under FINE WORK MEASUREMENT TO PRODUCE SYNTHETIC DATA in part 4 of this chapter.

Case 2—Work measurement of operations performed by a machine or process which is mainly under the control of an operator

This procedure is suitable for measuring operations by fine work measurement which are either repetitive or semi-repetitive in nature; which are performed with the aid of powered machinery under the control of the operator; and where cost reduction is to be achieved by using the work values for bonus incentive payment.

Typical examples include machining by drilling machine, capstan lathe, milling machine, etc.

STAGES 1–5

This would follow the same procedure as for case 1.

STAGE 6

As in case 1, this would consist mainly of time study except that it would be necessary to carry out the full procedure, including consultation with technical authorities on the machines and processes involved, and the taking of time studies on these processes (steps (*a*) (iii) and (*c*) (iv) respectively).

323

STAGE 7

The procedure would be the same at this stage as for case 1, except that in addition the machine process times should be analysed to establish possible relationships between these and the product or process, as explained in step (c) (i).

STAGES 8–12

The same remarks apply to these stages as in case 1.

REFERENCING PROCEDURE

The same referencing system is used as for case 1.

Case 3—Work measurement of repetitive or semi-repetitive operations performed by a team of operators

This procedure is suitable for measuring work of a repetitive or semi-repetitive nature performed by a team, where the cost reduction is to be achieved by time values used for bonus incentive payment produced by fine work measurement.

Typical examples include light assembly operations carried out by a team of operators working alongside a conveyor belt.

STAGES 1–6

These stages would be similar to that adopted in case 1.

STAGE 7

This would follow the same pattern as that used in case 1 except that the calculation and reduction of unoccupied time would be carried out as explained in steps (c) (i) and (iii) (reference Chapter 11).

STAGES 8–12

These would follow the same procedure as in case 1, except that a bonus incentive system specially designed for team workers may have to be considered (reference Chapter 14).

REFERENCING PROCEDURE

The same referencing procedure is used as for case 1.

Case 4—Work measurement of machine or process controlled operations

This process is suitable for the use of fine work measurement techniques to measure work on which operators are to be persuaded to ensure that automatic or semi-automatic machines or processes are maintained at a high utilization in return for an incentive payment—particularly where more than one machine is attended by an operator.

Examples occur on the operation of automatic turret lathes, automatic and semi-automatic plastic or rubber moulding machines, multi-slide presses, automatic looms, winding operations in the textile industry, etc.

STAGES 1–5

A close assessment of the machine utilization and the departmental productivity should be carried out at the survey stage, and the potential cost reduction from both increased productivity and increased recovery of fixed overheads. These are then set against any increases in pay necessary to persuade operators to produce at optimum performance to give the net benefits (reference Chapter 17).

STAGES 6 AND 7

The procedure is followed in its entirety and will consist mainly of time study, the reduction of unoccupied time and the minimization of interference losses (reference Chapter 10).

STAGES 8–12

These should follow the same general procedure as in case 1 except that a special incentive bonus system will have to be designed to ensure the highest possible machine utilization (reference Chapter 14).

REFERENCING PROCEDURE

The same referencing procedure is used as for case 1.

Case 5—Work measurement of operations of a unique non-recurring nature

This procedure applies to the use of fine work measurement to measure work on which there is no basis for comparison with any other work being performed within the organization. Its use is very limited, and would only apply to single operations such as cleaning, or setting of one piece of equipment where there is only one method of so doing, etc.

STAGES 1–5

Since the measurement of the operations will usually form part of a more extensive work measurement programme, these stages would probably have been covered in a similar manner to that described in case 1.

STAGE 6

This stage would be greatly simplified, and steps (*a*) (iv) and the whole of (*b*) and (*c*) (iii) and (iv) would be omitted.

STAGE 7
This stage would merely consist of the following steps only—(*e*) (iv) and (v), (*f*), (*g*) and (*h*).

STAGES 8–12
These would normally form part of a more extensive work measurement programme and therefore would not normally be considered as isolated operations.

REFERENCING PROCEDURE
The referencing procedure used in this case would be as that given in part 4 of this chapter under FINE WORK MEASUREMENT OF NON-RECURRING WORK.

Case 6—Work measurement of long cycle non-repetitive operations

This procedure applies to the measurement of work using broad work measurement techniques and which is of a fairly long duration, which does not tend to recur, but throughout which there are common bases for comparison. Typical examples occur on maintenance work, tool making, clerical and drawing office work, storekeeping, etc.

STAGES 1–5
The procedure for these stages would be similar to that used for case 1, except that a greater allowance would have to be made for the cost of subsequent maintenance by work study staff. Where the measurement of maintenance work is contemplated, however, consideration may also have to be given to the installation of planned preventive maintenance.

STAGES 6 AND 7
These stages would be similar to those used for case 1 except that the broad work measurement techniques described in Chapter 8 would have to be used.

STAGES 8–12
The remarks made in case 1 apply equally to this case, and real efforts should also be made to simplify the synthetic data by extending the limits of accuracy as widely as possible and so designing the incentive to minimize fluctuations in pay for relatively small fluctuations in effort.

The installation of labour cost control is equally as important to these applications as to repetitive work.

REFERENCING PROCEDURE

The procedure used for referencing should be that as described under BROAD WORK MEASUREMENT in part 4 of this chapter.

Case 7—Method study of operations of an extensive nature

This procedure applies to cost reduction on relatively large-scale operations such as workshop or office layouts, stores and warehouse layouts, internal transport problems, material flow problems, etc. using fine and broad method study techniques for their solution.

STAGES 1–5

During the survey stage, a close assessment would be made of possible reductions to labour cost, and this set against the cost of implementing and maintaining the revised methods to obtain the net benefits (reference Chapter 17).

STAGE 6

The operation carried out at this stage would mainly consist of recording information on standard forms preparatory to the construction of *process charts*, *flow diagrams*, *string diagrams*, etc. and the checking and cross-checking of these until satisfied that it is both authentic and complete. The full procedure as given would be followed except that steps (*b*) (iii), and steps (*c*) (iv) and (v) would not be carried out (reference Chapter 12).

STAGE 7

This stage consists of the detailed examination of any *process charts*, *flow charts*, *string diagrams*, etc. constructed to show the present methods, and the development of revised methods by analysis of these. Further charts and layouts would then be developed after experiment on a trial basis. The standard procedure would be followed except that the whole of steps (*a*) to (*e*) would be omitted (reference Chapter 12).

STAGE 8

This stage is the same as that given under the full procedure except that steps (*a*) and (*b*) would be omitted. If not already in existence, however, labour cost control should be installed at the earliest possible opportunity when work measurement is carried out.

STAGES 9–12

Maximum assistance should be given by the work study department to all the staff involved in the implementation and operation of the

327

revised methods to bring them to a successful conclusion within the original estimated time.

REFERENCING PROCEDURE

The referencing procedure to be used is as that described under FINE AND BROAD METHOD STUDY in part 4 of this chapter.

Case 8—Method study of operations of a concentrated nature

This procedure applies to cost reduction by using method study of repetitive operations performed alongside a conveyor belt, intricate hand operations such as soldering, thread manipulation in textiles, line inspection (i.e. 100 per cent inspection), light assembly operations, rearrangement of workplace layouts, etc. and bringing about improvements to these by simplifying the work movement pattern.

STAGES 1–5

During the survey stage, assessment would have to be made of possible reductions in labour cost due to reduction in work content brought about by the improved methods of operation. This would have to be balanced against the cost of the investigation, subsequent re-training programmes, installation and other costs where new equipment is involved (reference Chapter 17).

STAGE 6

The procedure at this stage would follow the same pattern as that set out in parts 2 and 3 of this chapter except that steps (c) (iv) and (v) would be omitted. It would consist of recording information either on standard forms, by ciné camera and *simo chart*, or by cyclegraph apparatus and still camera according to the degree of observation and the nature of the application (reference Chapter 12).

STAGE 7

The procedure adopted at this stage would be similar to that outlined in Case 7 except that examination of present methods would be carried out using *simo charts*, *cyclegraphs*, etc., and improved methods then developed by analysis and trial.

STAGE 8

This stage would consist mainly of drawing up plans for re-training, installation of improved workplace layouts, etc., using the standard procedure, but omitting steps (a) and (b).

If labour cost control is not in existence, it should be installed at the earliest opportunity after work measurement.

STAGES 9–11

The remarks under Case 7 apply equally well to this example.

REFERENCING PROCEDURE

The procedure used for referencing of data is described under MICRO-MOTION STUDY in part 4 of this chapter.

Case 9—Method study of material usage

This procedure applies to the reduction of material cost by investigation into material usage and the installation of some form of material waste control. Typical applications occur where extensive use is made of material during processing as in casting, metal extrusion, etc.; or where expensive material is used such a nickel, chromium, silver and gold, etc.; or where scrap is non-recoverable as in the manufacture of leather goods, rubber moulding, textiles, etc.

STAGES 1–5

During the survey stage, the actual amount of material waste taking place should be obtained from official records and an assessment made of the possible reductions. Evaluation of this is then made against the material cost to give the gross financial benefits. Against this is set the cost of implementing the revised methods, cost of equipment, cost of pay increases due to bonus incentive payments, etc. to give the net cost reduction.

STAGE 6

This would follow the full procedure except that steps (*b*) (iii) and (*c*) (iv) and (v) would be omitted. The data collected would consist mainly of detailed measurements of length, weight, area or volume of material used or wasted at each activity centre.

STAGE 7

This stage would consist of steps (*f*), (*g*) and (*h*), improved methods of operation being developed to reduce material wastage and these tested in practice before bringing them to finality.

STAGE 8

This would follow the same pattern as the full procedure except that it would not include the installation of labour cost control.

Since the reduction in material waste may have been achieved by alterations to operational methods, means of re-training labour and the introduction of a bonus incentive system may be necessary to reach and maintain higher standards of material utilization.

329

STAGES 9–12

New methods of operation are launched and the maximum assistance given to labour to ensure achievement and maintenance of the new material utilization standards set.

REFERENCING PROCEDURE

The procedure used for referencing of data should be as that described under FINE AND BROAD METHOD STUDY in part 4 of this chapter.

Case 10—Method study of clerical systems

This procedure applies to reduction in cost and improvement to information services by the detailed investigation and redesign of clerical systems.

Sometimes known as Organization and Methods (O and M), examples in this category would include investigation into systems of wage payment, cost control, etc., either with a view to simplification and increased effectiveness, or prior to the installation of data processing. The procedure can only be used to advantage with the assistance of specialists in the clerical techniques. This particularly applies to sales procedures, cost control, production control, or where computers are to be used.

STAGES 1–5

A very careful assessment of the implications of improving any of the clerical procedures will have to be made during the survey stage. For in the case of production control, fundamental changes to policies on stock control, material flow, etc. may be introduced which could have far-reaching consequences. This is why it is so important to consult specialists in the particular techniques throughout.

Streamlining of clerical methods without alteration to the basic principles, however, can often be achieved with resulting reduction in clerical work. This should be assessed against the background of investigation and installation cost to arrive at a net cost reduction.

STAGE 6

This would generally follow the standard procedure except that steps (*c*) (iv) and (v) would be omitted. It would consist mainly of collecting data by questioning and recording these observations on standard sheets. From this information document flow charts would be constructed and checked with the originating information until their accuracy was ensured (reference Chapter 12).

330

STAGE 7

This consists of the detailed examination of the document flow chart, and of the documents themselves, with a view to reducing duplication of effort and preventing inessential information being recorded, or information being sent to the wrong destinations. Revised methods are then developed to reduce clerical time and effort to a minimum whilst retaining the essential characteristics of the information system (reference Chapter 12).

STAGE 8

This would consist of steps (*c*) and (*d*) in the standard procedure, during which a complete plan of operation for the new conditions would be drawn up, and the net cost reduction calculated.

STAGES 9–12

When introducing new clerical methods, a great deal of experimenting should be carried out at this stage with altered clerical forms produced by a stencil and in limited supply. In the light of knowledge gained, the final formats can then be produced as a printed form.

Case 11—Investigation into the feasibility of installing newer types of machines or processes

This procedure can be used to test the feasibility of proposals made to install a type of machine or process different to that in present use.

STAGES 1–5

The survey would consist of obtaining reliable evidence of the operating speeds of the proposed equipment—preferably from direct observation at a trial demonstration. Other details which would have to be considered are setting times, quality standards of the product and the capital cost of the equipment including any tools. Technical staff should be actively consulted also to ascertain the functional reliability of the equipment.

From these details, an assessment should be made of the possible reduction in production costs, and those set against the capital cost of the machine, tools, and the installation and maintenance costs in order to test the feasibility of its installation.

STAGE 6

This would take the form of a more prolonged trial run of the proposed equipment under simulated conditions of actual operation, the procedure being confined to step (*c*) (iv) only.

331

Summary of procedures

STAGE 7

The whole of this stage is reduced to the one single step (*h*) which is the determination of a reference period of existing methods.

STAGE 8

This would be confined to steps (*c*) and (*d*) only, during which final recommendations would be drawn up, balancing the cost reduction against the installation costs.

STAGE 9

During this stage final recommendation to management would be made as to whether to proceed or not, depending on whether the capital cost of the equipment could be recovered by the net cost reduction within two years.

STAGES 10–12

If the project is sanctioned, work study staff should make detailed investigations to ensure that the benefits originally claimed are met.

Case 12—Work measurement using predetermined motion-time systems

The procedure can be used to accommodate the application of any of the predetermined motion-time systems (methods-time measurement, Work Factor, etc.), the elements of work being built up in exactly the same way as when using time study or broad work measurement with the aid of a time piece (reference Chapter 11).

19

Other techniques

Work study is not the only cost reduction technique. For example, if work measurement-based bonus incentive payment systems are installed in areas where work planning methods are in need of improvement, they will only be partially successful and may even fail altogether, particularly if operators are prevented from earning a reasonable bonus by being continually obliged to wait for work for long periods.

In these situations, it is necessary to combine work study with other related techniques. Some of these are now briefly described.

Production planning and control

Production control can be described as "the control of production from the time when the order for the product or service is received, until it is delivered to the customer's premises, having maintained balanced stocks and work-in-progress at the minimum level."

The design and operation of a successful production control system can be a complex and highly specialized undertaking. Like work study, it should never be attempted by anyone who is not properly qualified to understand it.

Poor control systems inevitably result in too much of certain types of material and components in stock and not enough of others. Smooth work flow is often restricted by excessive amounts of material on the shop floor. Lack of central control may also mean inadequate instructions being issued. All these will add up to operators having to wait for work, and prevent high performance levels. This will again lead to dissatisfaction and aggravate industrial relations. Introducing work study in these conditions will only be a

333

further irritant which may lead to a complete breakdown of industrial relations. Under these circumstances the wisest course is to improve the control system first and introduce work study afterwards—or combine the two together.

Overhaul of a production control system involves the following—

(a) The determination of a policy of production, selling, and finance which are balanced one with the other. This is the function of the governing board, or board of directors.

(b) The rationalization of the management structure to define the functions of sales and production.

(c) The determination of a sales requirement schedule by analysis of demands for the products of the organization using sales forecasting techniques and market research, where these are appropriate.

(d) The creation of a production schedule which exactly reflects the sales demands but which is also a realistic possibility in terms of labour and machine capacity.

(e) The arrangement of stores and adequate but not excessive levels of stocks of raw materials, purchased and manufactured parts, ancillary materials and tools to meet the requirements of the production schedule.

(f) The creation and regular issue of periodic programmes of work to each department, which are within the capacity of those departments, and for which adequate provision has been made for the supply of any necessary material and parts within the period specified.

(g) The formulation of a rational progress system which clearly indicates priorities to all concerned.

(h) The control of distribution of finished goods and services to customers in accordance with the sales requirement schedule.

It is important to understand the difference between a programme and a schedule. A programme is a list of articles or work which is required within a specified period (a week or a month, etc.). A schedule is a series of lists (or programmes) arranged on one form. Usually it is headed with week numbers, period numbers or months, the quantities of each of the goods listed on the left-hand side of it, being shown under the appropriate period when it is wanted. Listing any one week, period or month will then produce a programme for that period.

Project network analysis
This is a system of planning which can be used for such projects as building a house, building a boat, publishing a book, launching a

new product, carrying out major repairs to plant during an annual "shut down" period, etc.

It employs a simple symbolism to represent jobs, known as *arrows*, the beginning and ends of them being indicated by circles called *nodes*—

Node   Arrow   Node

Fig 70 Project network analysis symbols

The arrow can be of any length and need not bear any relationship to the relative length of time which each job takes.

Where there are several jobs which form a complete project, some of which can be carried out at the same time as others, nodes can be linked together by arrows to form a *network*. In this case it is usual to number the nodes for reference and indicate with the duration of each job over the arrow. Where nodes need to be linked together to indicate continuity and there is no job to be done between them, a *dummy arrow* can be used. This is represented by a dotted line marked with an arrow.

Various paths can be traced out through a network giving varying lengths of total time to complete the project. The path which will give the longest time in this way is the shortest time in which the job can be done, since any other person doing jobs of shorter duration at the same time would have to wait while the longer one was finished.

The path traced out by these longest jobs is known as the *critical path* and the jobs themselves as *critical jobs*. The enforced waiting time due to jobs of varying lengths being carried out simultaneously is known as *float*.

The length of some jobs used in building networks can be determined from synthetics or predetermined motion-time data. Other work, such as designing or planning, etc., has to be estimated using statistical formulae, an example of which is given in Chapter 8. Where these are used, the technique is sometimes called PERT (Program Evaluation Review Technique). Special programs have been written to enable this type of planning to be handled by a computer. In such cases valuable information for the allocation of resources of various types of labour can be quickly obtained, enabling waiting time to be minimized and the complete project to be completed within the planned cycle time.

The advantage of using a computer for PERT planning is that if details of completed jobs are fed into it as the work proceeds, it can be programmed to print out revised schedules showing different critical jobs brought about by non-adherence to the plan so far.

Process planning

Process planning is in reality a work study technique, although it is not generally considered as such. It is one of the major links which join work study to production control.

It is basically a technique of dividing work into separate operations which are set down on an instruction sheet known as an operation layout or planning sheet. Process planning engineers are mainly specialist technicians, because their function is to decide the most suitable manufacturing techniques, the correct use of machines, the form and general design of tools and other special purpose equipment. But their work can be made much more effective if technical knowledge is combined with work study principles. That is why the process planning section should form part of the same administrative structure as the work study department.

Operation layouts are the medium through which instructions are translated via the drawings and other specifications to the shop floor, and therefore to be complete, standard times should be added to them to form a complete record of such operations. Layouts containing such data not only provide a means of issuing standard times to the foremen and operators, but also form a valuable basis for costing.

Variety reduction

Often, examination of a company's products reveals that the bulk of them consist of a relatively small number of different items, whilst the remainder are made up of a greater—and quite often excessive—variety, which are also in far less demand. This can make for unnecessary complex conditions in production, documentation and control. Rationalization of such a situation to reduce variety by withdrawing certain designs which have been obsolete or reached a minimal demand can, sometimes, bring about considerable economies.

There is no set procedure for this technique—it merely consists of the statistical collection of information followed by a critical examination. To be really effective, it should be carried out in full consultation with design, manufacture and sales functions, as withdrawal of a product in small demand from a valuable customer without proper investigation can lead to loss of customer good will.

Value analysis

Value analysis is the critical examination of *functions* to determine whether the same *functional* product can be manufactured at lower cost. In other words, it questions each particular part, material, method of manufacture, source of purchasing, and other factors which contribute to the cost of a product; and attempts to discover methods of producing a product at lower cost with no reductions in quality, safety features or aesthetic appeal.

One way of carrying out this technique is to form a committee who are encouraged to make any suggestions on different manufacturing and design methods, however unusual or unorthodox they may seem, and then to evolve from these some means of achieving the same end result with less overall cost.

Bibliography

The following is a list of books recommended for further reading—

Work Study by R. M. Currie (Pitman, London).
A comprehensive guide to the whole subject.

Introduction to Work Study (International Labour Office, Geneva).
An excellent introduction, particularly for Method Study.

Glossary of Terms Used in Work Study (British Standards Institution, London).
Invaluable for the student and practitioner as a guide to standard terminology.

The Purpose and Practice of Motion Study by Anne Shaw (Columbine, London).
An excellent book on the subject by a world authority.

Methods-Time Measurement by Maynard, Stegemerten and Schwab (McGraw-Hill, New York).
The standard textbook on basic MTM by the originators of the system.

Simplified PMTS by R. M. Currie (BIM, London).
A self-teaching manual on a second generation predetermined motion-time system developed at ICI. Recommended to the student and practitioner.

Industrial Engineering Handbook Edited by H. B. Maynard (McGraw-Hill, New York).
A comprehensive handbook to management techniques. Contains descriptions of many work study techniques including work measurement and predetermined motion-time systems.

Production Handbook Edited by L. P. Alford and J. R. Bangs (Ronald, New York).
An excellent handbook to industrial engineering techniques.

338

Office Organization and Method by G. Mills and O. Standingford (Pitman, London).
A most comprehensive and clearly written guide to office machinery and O and M. Strongly recommended.

Cost Accounts by Walter W. Bigg.
One of the best textbooks ever written on this subject.

The Principles of Production Control by John L. Burbidge (Macdonald and Evans, London).
A thoroughly recommended manual written by a master of this subject.

Network Analysis by Albert Battersby (Macmillan, London).
Recommended for the student and practitioner.

Technique of Value Analysis and Engineering by Lawrence Miles (McGraw-Hill, New York).
An excellent practical manual on the subject.

Facts from Figures by M. J. Moroney (Penguin Books).
A clearly written and comprehensive guide to statistics invaluable to students, practitioners and management.

Glossary of terms

The following is a brief glossary of terms used. Wherever possible, terminologies have been based on those given in the "Glossary of Terms used in Work Study" published by the British Standards Institution (B.S. 3138:1969).

ACTIVITY SAMPLING
A technique in which a number of instantaneous observations are made of a particular operation from which the ratio of the activities and non-activities to the total time can be calculated.

ACTUAL COST PER STANDARD HOUR
Gross wages divided by useful work in terms of measured plus uncontrolled work at an assessed performance.

ALLOCATED WORK
Work for which a value has been given by allocating a number of actual working hours to it. It need not have been determined accurately.

ALLOWED TIME
A time obtained by increasing the standard time by an increment (usually one-third). It used to be considered that a piece worker would work at standard time speed, and a day worker at allowed time speed—the one being one-third faster than the other. This concept is no longer accepted as valid, however.

ANALYTICAL ESTIMATING
A method of estimating standard (or allowed) times by synthesizing them from individual elements, most of which have themselves been estimated from practical experience in the type of work.

ANALYTICAL OBSERVATION
A method of measuring work by intermittent observation using an observer who is trained in rating and is also familiar with the type of work.

340

ANCILLARY WORK
A service (such as setting, work preparation, etc.) which cannot be classed as productive, but which is ancillary to it. It is often carried out by direct workers.

ATTENDANCE TIME
The total time spent by a worker at the place of employment. Sometimes known as clock time or clock hours.

BASIC TIME
The net time taken to carry out work at standard rating (i.e. at BSI 100).

BREAK-EVEN CHART
A chart which shows graphically the estimated total expenditure and income, the point at which they cross being known as a break-even point.

BREAK POINT
The instant at which a work element ends.

CHECK TIME
The time between starting a time study and the start of observation, and between the finish of observation and the end of the study.

CLOCK HOURS
See attendance time.

CONSTANT ELEMENT
A work element whose basic time is constant.

DEPARTMENT PERFORMANCE
The ratio of total measured work to gross attendance hours. Sometimes, it may be decided to exclude from the attendance hours allocated work diverted time and waiting time for which the department is not responsible, and to add uncontrolled work at an assessed performance to the standard hours to give a gross measured hours value. In this case, the department performance is the ratio of the gross measured hours to the gross attendance hours less allocated time, waiting and diverted time for which the department is not responsible.

DIRECT WORK
Work which directly changes the shape, form, standard of finish or classification of a saleable item.

DIVERTED TIME
The preferred term to describe lost time (see lost time).

EFFECTIVE TIME
That portion of the elapsed time during which a worker is engaged in the proper performance of prescribed work. It is often referred to as net effective time in this book to emphasize its character.

ELAPSED TIME
The time from the start to the finish of a time study.

ELEMENT
A convenient portion into which work is divided for work measurement and analysis.

Glossary of terms

EXCESS COST
The difference between standard and actual cost. It can be expressed as a gross figure, or per standard hour.

EXCESS WORK
Extra work over and above that specified by control standards.

EXTENSION
The conversion of observed time to basic time.

FATIGUE ALLOWANCE
That part of the relaxation allowance necessary for a worker to take compensatory rest.

FIXED OVERHEADS
Overheads which do not tend to vary in amount over a financial period.

FLOW DIAGRAM
A diagram drawn to a distance scale to show the progression of work in space. Sometimes also known as a flow chart.

FLOW PROCESS CHART
A process chart which shows the flow of a product or work, recording the events using process chart symbols.

FLYBACK TIMING
A method of timing by stop watch where the hands are returned to zero and instantly restarted at the end of each element, these individual times being recorded on an observation form.

GOVERNING TIME
The time which controls the output of a group of workers, machines or processes. Sometimes known as lead time also.

INDIRECT WORK
Work which is essential to production but which only contributes indirectly to it, e.g. labouring, office cleaning, etc. It is usually not carried out by direct workers.

INEFFECTIVE TIME
That portion of the study time which is spent on an activity which is not a specified part of the job.

INSIDE WORK
Work which can be performed by a worker within the process-controlled (or machine-controlled) time.

INTERFERENCE TIME
That part of productive time which is lost due to more than one machine, process or operation ceasing its operation cycle before others can be restarted. It can occur in process-controlled or team work. Sometimes known as synchronization time.

LABOUR COST CONTROL
A method of presenting deviations from standard cost, performance and other factors to be used as a management control. It is termed "Labour control" in B.S. 3138:1969.

342

LEARNER ALLOWANCE

A temporary benefit given to workers under incentive until they develop ability.

LOST TIME

That part of attendance time when the worker is engaged in other activities, e.g. committee work, accidents, etc. The preferred term is now *diverted time*.

MACHINE UTILIZATION INDEX

The ratio of the machine running time at standard to the actual machine running time.

MAKE-UP

The amount of adjustment necessary to bring a worker's earnings to his guaranteed minimum when he has fallen below the performance level which earns that guaranteed minimum. Sometimes known as *below basic*.

MEASURED WORK

Work for which controlled standards have been determined by work measurement techniques.

METHOD STUDY

The recording and subsequent analysis of existing methods of work which lead to the development of more effective methods at reduced cost.

METHODS-TIME MEASUREMENT (*MTM*)

A series of predetermined motion-time systems originally developed from "therbligs". Three levels currently exist, *MTM*-1 (or basic *MTM*), *MTM*-2 and *MTM*-3, these being developed to measure short cycle, medium cycle and long cycle work respectively.

MICROMOTION STUDY

A term sometimes used to describe the analysis of movements by breaking these down into "therbligs".

MULTIPLE ACTIVITY CHART

A chart representing several inter-related activities on a common time scale.

OCCASIONAL ELEMENT

An element which occurs irregularly with respect to the work cycle or job.

OPERATOR PERFORMANCE

The ratio of the total measured work in standard time to the actual time spent working on this measured work.

ORGANIZATION AND METHOD (O & M)

A means of investigating and improving procedure; methods and systems, communications and controls, and organization structure.

OUTSIDE WORK

Work which can only be performed by a worker outside the machine- (or process-) controlled time.

OVERALL PERFORMANCE

The ratio of total measured work plus uncontrolled work at assessed performance to the gross attendance hours less allocated work.

Glossary of terms

PERSONAL NEEDS ALLOWANCE
That part of the relaxation allowance necessary for the worker to attend to personal needs.

PREDETERMINED MOTION-TIME SYSTEM (*PMTS*)
A technique of building up standard times for jobs using times determined for basic movements of the human body.

PROCESS ALLOWANCE
See unoccupied time allowance.

PROCESS CHART
Chart in which a sequence of events is shown diagrammatically using process chart symbols.

PROCESS-CONTROLLED TIME
The time taken for a machine or process to complete that part of a work cycle which is peculiar to the machine or process. Also known as machine-controlled time.

RATED ACTIVITY SAMPLING
A form of activity sampling where rating is applied to enable the work content of the operation to be determined.

RATING
A method of indicating the rate at which work is being performed by using numerical values.

RELAXATION ALLOWANCE
An allowance which is added to basic times to provide the worker with the opportunity to recover from the effects of fatigue and give attention to personal needs.

SIMULTANEOUS MOTION CYCLE CHART (SIMO CHART)
A chart which records "therbligs" on a motion-time scale—usually constructed from analysis of ciné film.

STANDARD COST
The cost of a unit of work at standard performance under standard conditions of pay and working hours. It is usually expressed in terms of direct, indirect and total cost per standard hour, standard minute, etc.

STANDARD PERFORMANCE
The rate of output which qualified workers will naturally achieve over a working day with reasonable incentive and without over-exertion, but with the adequate relaxation needed for the work.

STANDARD TIME
The time in which a job should be completed at standard performance.

STANDARD UNIT OF WORK (OR WORK UNIT)
A fraction of work at standard performance plus a fraction of relaxation, the proportions of which should vary according to strain but the total always aggregating unity. It should be so set as to allow a qualified worker of average ability to produce the same number of standard units of work as his attendance time, providing he is not restricted in any way from so doing.

344

STRING DIAGRAM
A diagram on which a thread or string is used to trace out the path of movement of workers, materials or equipment.

STUDY TIME
The elapsed time of a time study minus the check times.

SYNTHETIC DATA
Data derived by work measurement which will enable standard times to be built up for complete operations by synthesis.

TAPE DATA ANALYSIS
A method of recording work measurement data using a portable tape recorder.

TEAM WORK
Work done by a number of workers in close association one with the other.

THERBLIG
A term originated by Frank B. Gilbreth to denote the absolute division of movements of the human body in performing work.

TIME-MEASUREMENT UNIT (*TMU*)
A unit of time used by some of the predetermined motion-time systems which is equivalent to one-hundred thousandth of an hour (0·00001 hours).

TIME STUDY
A work measurement technique used for recording, measuring and analysing work in terms of elements by using a time piece (i.e. stop watch or wrist watch).

UNCONTROLLED WORK
Work for which no standards have been determined. It is sometimes referred to as unmeasured work.

UNOCCUPIED TIME
A period where workers are obliged to cease work to allow a process, machine or other operator cycle to take place when all the conditions are at standard performance.

VARIABLE ELEMENT
An element for which the basic time varies in relation to some characteristics of the product, process or equipment.

WAITING TIME
That part of attendance time (other than unoccupied time) during which the worker is obliged to wait due to circumstances beyond his control.

WORK FACTOR
A predetermined motion-time system.

WORK MEASUREMENT
A method of determining times that a qualified worker needs to carry out a specified job at a specified level of performance.

WORK SPECIFICATION
A specification of the precise methods used to perform a job in order to qualify issued standard times.

WORK UNIT
See standard unit of work.

Index

Abbreviated Work Factor 221, 232
Accuracy of standard times 92, 179–182, 238
  checking 85–94
Activity sampling 2, 96, **101–16**
  accuracy 108, 110, 115
  formulae 108, 113, 115, 116
  number of observations 108, 110, 115
  of process-controlled work 164
  procedures 110–16, 318–19
  referencing 318–21
Actual costs 266, 268, 340
Adaptations of work study procedure 322–32
Allocated work 340
Allowances 263
  learning 183, 258, 259
  policy 263
Allowed time 29, 30, 340
Analytical estimating 2, 95, 124, 340
  procedure 99–101
  referencing 318–21
Analytical observation 96, 340
  procedure 117–23
Ancillary allowance 163–4
Ancillary work 263, 268, 341
Arithmetic mean 21–4
Assessment of potential benefits 299–300
Attendance time 242, 245, 263, **265**, 268, **341**
Averaged expected time 99
Average rating—see rating

Base wage 244–6
Basic times—(see also elements) 40, 43, 71, 91, 341
  constant 50–3, 72–5, 225

selection of 47–59
  variable 53–8
Bar charts (see multiple activity charts)
Benefits, assessment of 299–300
Bonus earnings 244, 248, 249
  incentive (see incentive bonus)
  rate 246
  tables 247, 251, 254, 256, 258
Break-even chart 272, **273**, 341
Break point 36, 341
British Standard Institution rating system 14, 15, 19, 21, 29, 32, 225
Broad analysis techniques 2
Broad method study 2, 140
  referencing procedure 321–2
Broad work measurement 3, 95–124
  referencing procedure 318–21
Budgetary control 277

Centiminutes **31**, 37, 43, 53, 78–80
Check studies 88, 89
Check time 37, 38, 40, 43, 90, 165, 341
Checking—
  rating 28
  relaxation allowance 94
  standard times 88–94, 145–8, 309–11
Chronocyclegraphs 197, 209
Ciné-equipment, use of 193–4, 207–9
Clerical staff in work study department 8
Clerical work—
  measurement of 97, 107–9, 111–12
  procedures recording 190–1
  routines, improving 215
  study of 190–1, 330, 331
Clock hours—see attendance time
Collection of elements sheet 76, **77**, **81**, 145, 316, 317

347

Index

Computation—
  form for relaxation allowance   68
  form for synthetics   177, 178
  of relaxation allowances   60-1, 67-8
  of standard times from synthetics   85-87
  of standard times from time studies   86-8
Concept of standard effort   18-20
Constant elements   78-80, 176, 341
  segregating from variable elements   49
Contingency allowance   71
  determination of   75-8
Contract costing   274
Control sheets for labour cost   269
Cost—
  factory   274
  labour   159, 162, 270, **271**, 275, 276
  marginal   274
  material   270, 271, 275-6
  overhead   270, 271, 275-6
  per machine hour   159, 162
  prime   274
  product, determination of   275-6
  production   275
  reduction reports   290-2, 313-14
  sales   275
  standard   344
  total   159, 163, 171, 276
Costing   270, 274
  contract   274
  department   274
  job   274
  operation   274
  process   274
  standard   274
Critical path   335
Currie, Russel   227, 232
Cyclegraphs   197, 209

Data blocks   238
Data sheets   316-22
  used for flow charts   204, 208
  used for multiple activity charts on process-controlled work   138, 150, 152, 155
  used for multiple activity charts on team work   127, 132, 134
  used for process charts   202-4
  used for process time study   142
  used for synthetic data   87
  used for workshop layout   45, 134
Department costing   274
Department performance   148, **264**, 268, 279, 280, 283, 287, 300, **341**
Detailed Work Factor   220-3
Differential output incentives   246-9
Differential piecework   246-7
Direct labour reference period   278-80
Direct standard cost   265-6

Direct work   262, 341
Directly based incentives for indirects   256, 257
Directly proportional incentives   241-6
Diverted time (see lost time)
Division of work into elements   20, 96-7
Documents
  analysis of   197, **209**, **212**
  method study   7
  predetermined motion-time systems   7
  time study   7
  work measurement summary   7

Effective operator performance   166, 170
Effective time   **40**, **42**, 43, 90, 91, 103, 104, 106, 165, **341**
Elapsed time   **40**, **43**, 90, 103, 104, 106, 165, **341**
Elements   **35-7**, 43, 47-59, 71, **341**
  classifying   74-5
  constant   78, 79, 80, 341
  constant, selection of   50-3, 72-5
  division of work into   **20**, **21**, 25, 49, 50, 96, 97
  ideal length of   37
  irregular   36
  occasional   36, 343
  regular   36
  repetitive   36
  variable   78, 79, 176
  variable, selection of   53-8, 170
Element registers   72
Equipment—
  recording the movement of   189-90
  work study   6, 7, 31, 32
Estimating   99
  analytical   2, 95, 114, 340
  analytical, procedure   99-101
  analytical, referencing   318-21
  time for work measurement   184-5
European rating system   19
Examples of incentives—
  inspection of pressure gauges   256
  machine utilization in textiles   254
  reducing scrap on watch straps   225-6
  supervision   257-8
Examples of method study—
  machining on a lathe   137-9
  making tea   126-8
  on a "flow-line" team   131-6
  on a team of two   131-2
  waste reduction on extrusion processes   216-17
Examples of work measurement—
  activity sampling on multi-machining   106
  assembly of fountain pen   234
  clerical work   107-9, 111-12
  drilling   43

348

Examples of work measurement—*contd.*
machining by capstan lathe 177–8
painting by maintenance workers 100–6, 118–23
setting on capstan lathes 177–8
spray painting clock dials 45–6
winding in the textile industry 139–72
Excess cost 266, 268, 269, 342
Excess work 263, 342
Extension of observed times 15, **21–8**, 342
Extreme values 27–8

Factory cost 274
Fatigue allowance 60, 342
Fine method study, referencing 321
Fine work measurement, referencing 315–18
Fixed overheads 272, 273, 342
Fixed overhead recovery 277
Float 335
Flow charts 3, 197, 342
constructing 205
three-dimensional 208
two-dimensional 208
Flow diagrams 2, 204, 342
constructing 205
Fly-back timing 342
Frequency 82–4
Fundamentals of work measurement 14–30

Geared incentives 242–3, 247–8
Gilbreth, Frank and Lilian 4, 194
Governing—
operation 130, 133
time 130, 342
Graph sheet 320
for frequency comparison, 83, 84, 145
for multiple activity charts 317
for simo charts 211
for variable elements 56, 57
Gross earnings 244–6
Gross time values 130, 133, 135
Group incentives 253
Group technology 215
Guaranteed minimum wage 242, 244–6

Handle 237
Hours
clock 265
machine 162–3
standard 82, 133, 135

Idle machine time 152
Idle time 90, 165
Improvements to—
clerical routines 215
manufacturing methods 214–15

material flow 212–13
operating conditions 213–14
Incentive bonus systems 240–61
choice of 259
differential output 246–9
differential piecework 246–9
direct labour 241
directly proportional 242–6
earnings from 244, 248, 249
group 253
improvements to quality 255
indirect labour 256–7
indirectly based 257
inspection 256
launching 260, 312
limiting differential 248–50
machine utilization 252–4
measured daywork 249, 251
modified standard time 247–8
piecework 246–7
premium pay plan 251–3
rate of pay 246
reduction of material waste 255–6
revision of 261
special 253–6
stabilized output 249
standard time 242–6
supervision 257–8
table of values, 247, 251, 254, 256, 258
work measurement based 4
Independent time piece 31
Indirect—
excess cost 267
labour reference period 280–1
standard cost 265–6
work 342
Indirectly based output incentives 257
Ineffective time 37, 40, 43, 91, 103, 104, 342
Inside work **137–9**, 146, 147, 150, 151, 342
per cycle 148, 150–2, 156–8
Integrated Standard Data (ISD) 232
Interference—
allowance, calculation of **157–72**, 309
losses on process-controlled work 342
losses on team work 136, 342
Interruption, the effect of 183–4
Investigation procedure 295, 301–5
Irregular occurring elements 36

Job card 176
Job costing 274
Job enlargement 240–1

Labour control (see labour cost control)
Labour, convincing 5
Labour cost 271, 273, 275, 276
Labour cost control 4, 15, **262–9**, 312, 342

Labour cost per machine hour 159, 162
Labour utilization index 159, 161, 162
Learning allowance 258, 259, 343
Learning, the effect of 183
Load factor 138, 149
Longest cycle time 153, 154
Lost time 267, 343

Machine—
  controlled work (see process-controlled
    work)
  hour, labour cost per 159, 162
  hour, total cost per 157, 162
  interference 157–72
  operating conditions, checking 139–
    144
  output, calculating 167–9
  utilization index **149**, 153, 154, 156,
    159, 161, 166, 168, 254, **343**
  utilization index reference period 281
Machines—
  investigating new types of 331–2
  minimum number allocated 158
Maintenance work, measurement of
    101–7, 117–22
Make-up 267, 343
Management by exception 215
Management, convincing 4
Manual work produced 168, 169
Manufacturing methods, improving
    214–5
Marginal cost 274
Master Clerical Data (MCD) 238
Master Standard Data (MSD) 227
Material—
  audits and analyses 197, 212
  cost 271, 273, 275, 276
  recording the movement of 189, 190
  wastage, recording 195, 196
  wastage, reducing 216, 217
Measured daywork 249, 251
Measured work (see also standard
    times) 92, 166, 263, 268, 280, 343
  checking the accuracy of 85–94
Memotion study 193
Method study
  analysis techniques 212
  broad 2, 140
  collection techniques 188–9
  consolidation techniques 218
  development phase 217
  documentation 7
  equipment 7
  fine 3
  micro 3–4
  of clerical work 190–1, 330–1
  of equipment 189–90
  of machine-controlled work 192–3
  of material movement 189–90
  of material usage 329

  of operator's movements 191–5
  of path of movement 195
  of plating processes 196
  of team work 192–3
  of work operations 327–9
  preparation for installation of 217–18
  preparation for investigation of 187
    188
  presentation techniques 197
  procedures 187
  referencing procedures 321–2
Methods—time measurement (MTM)
    219–20, 224–38, 343
  MTM—1 224–33, 235, 343
  MTM—2 122, 232, **234**–7, 343
  MTM—3 237–8, 343
Micro-analysis techniques 2, 3
Micro-motion study 343
  recording for 193–4
  referencing procedure 321–2
Middle Minute Data (MMD) 227
Minimum relaxation allowance 61, 67
Minimum wage 244–6
Modal average 24, 25
Modifications to product design 215–16
Modular Arrangement of Pre-determined
    Time Standards (MODAPTS) 232
Motion analysis sheet 210
Motion study (see also micro-motion
    study) 3, 193–4
Multiple activity charts 2, 3, 125, 343
  on process-controlled work 125–8,
    137–8, 150–2, 155
  on team work 131–4
Multiple machine work 151–72

Net effective time (see effective time)
Net measured work produced 166
Net standard times for process-controlled
    work 145–9
Net study time 165
Net time values 130, 133, 135
Net unoccupied time 153, 154, 156
Network analysis (see project network
    analysis)

Occasional elements 343
Office work, measurement of 97, 107–9,
    111–12
Operating conditions, improvements to
    213–14
Operation costing 274
Operation efficiency 131, 136
Operation layouts 11, 176, 336
Operator performance 93, 246, 248,
    250–2, 256, 259, **263**–4, **343**
  effective 166
Operator work cycle time 148, 151–4
Operators, recording the movement of
    192–3

Operators, recording work done by 191–2
Organization and Methods (O and M) 9, 343
Output costing 274
Outside work 137–9, 146, 147, 343
    per cycle 148, 150, 152, 156–8
Overall performance 264, 343
Overhead cost 270–3, 275, 276
    fixed 272, 273, 277, 342
    semi-variable 272
    total 273
    variable 272
Overhead recovery accounts 276–7
Overtime premium 268–9

Performance—
    and rating 91, 92, 265
    department 148, **264**, 268, 279, 280, 283, 287, 300, **341**
    during reference period 279, 288
    effective operator 166
    operator 93, 246, 248, 250–2, 256, 259, **263**, **264**, **343**
    overall 264, 343
    standard 18–20
Personal needs allowance 60, 344
Personnel department 11, 13
Piecework 242–4
    differential 246–7
Planned preventive maintenance 4
Policy allowance 263
Premium pay plan incentive 251–3
Pre-determined motion-time systems (PMTS) 2, 3, 6, 96, 122, **219–39**, 294, 344
    analysis sheet 234
    documentation 7
    equipment 7
Prime cost 274
Primary Standard Data (PSD) 227
Procedure in detail 298–314
Procedure in outline 294–8
Process allowance (see also unoccupied time) 125, 344
Process analysis 212
Process chart sheet 197, **199**, 321
Process charts 197
    constructing 200–5
    multiple 201–3
    preparing data for constructing 197
    representing movement 201, 204, 205
    simple 201–2
    symbols 197–9
    two-handed 203
Process-controlled time 137–9, 142, 344
    per cycle 148, 152, 153, 156–9
Process-controlled work, measurement of **137–72**, 324–5
Process costing 274

Process planning 2, 11, 336
Process time study 141–4
Program Evaluation Review Technique (PERT) 101, 335
Product cost 275–6
Product design, improvements to 215–16
Production control (see production planning and control)
Production cost 275
Production planning and control 4, 11, 333–4
Production studies 40, 88–91, 164–5
Project network analysis 2, 101, 334–6

Random numbers 114, 179
Random observations 101–15
Rate of producing on machines—
    actual 168
    at 100% utilization 167
Rated activity sampling **116–17**, 344
    on process-controlled work 165
Rating 91, 344
    average 42, 43, 90, 91, 93, 105, 165
    British Standard system (BSI) 14, 15, **19**, 21, 29, 32
    demonstration of 15–18
    European system 19
    60/80 system **19**, 21, 29
    100/133 system **19**, 21, 29
Ready Work Factor **221**, 232
Reciprate graph 25, 26
Reference period **278–82**, 311
    direct-labour 278–80
    inclusive of indirect labour 280–1
    inclusive of overheads 281
    machine utilization 282
    material utilization 281–2
    method study 282
    performance 279–88
Referencing procedure 315–22
Register of constant elements sheet 71–4, 88, 89, 100, 316–20
Regular elements 36
Relaxation allowance 37, 40, **60–70**, 91, 92, 317, 344
    adding 72
    computing 60–1, 67–8
    extra for females 67
    minimum 67
    outdoor work 69
    overall percentage factor 165–6
    tea breaks 69, 70
Relaxation allowance computation sheet 68
Repetitive elements 36
Reporting on recommendations 301
Reports—
    cost reduction 290–2, 313–14
    survey 285–7

Index

Sales cost 275
Savings calculations 287–9
Second generation motion-time systems 227, 232–4, 235–7
Semi-variable overheads 272
Set-up register **49**, 321
Simplified Pre-determined Motion-Time System (SPMTS) 122, 227, 232
Simplified Work Factor 221, 232
Simultaneous motion (SIMO) charts 3, 207, 209, **211**, 214
  symbols 206–7
Special incentives 253–6
Specialized motion-time data systems 238–9
Stabilized output incentives 249–53
Standard costing 274
Standard costs 265–6
Standard effort 18–20
Standard hours produced 245, 246
Standard performance 18–20
Standard rating 18–20
Standard Sewing Data (SSD) 239
Standard times 29, 30, 71, 75, 76, 78–82, 90, 92, 249, 263, 344
  accuracy of 179
  calculation of elemental 78, 79
  checking 88–94, 145–8, 309–11
  computation from synthetics 85–7, 176
  computation from time studies 86–7
  determination of final 171–2
  determination of net 145–7
  inclusive of interference allowance 171–2
  issue of 174
  net, per unit of output 149
  register of 182–3
  table of expected accuracies 180
Standard unit of work (see work unit)
Standards data, referencing 176–7
Stopwatch 31
String diagrams 3, 197, 205, 345
Study board 31
Study register **39–40**, 197
Study summary—constant elements sheet 52, 72, 315–20
Study summary and register of variable elements sheet 54, 55, 74, 316, 317, 319, 320
Study time 165
Summary of procedures 293–332
Supervision, convincing 5
Survey performance 283
Survey reports 285–7, 295, 301
Surveys 282–5, 295, 298, 299
Synchronization (see interference)
Synthetic data 176–8
  compilation of basic 71–84
  computation forms 177–8

Tape data analysis 239, 346
Tape recorder used for work measurement 122, 239, 345
Tea breaks 69–70
Team work, measurement of **125–36**, 345
Team work, method study of 37, 38
Textile industry, work measurement in 139–43
Therbligs 4, 194, **206–7**, 224, 345
Third generation motion-time system 237, 238
Three-dimensional flow charts 208
Time—measurement unit (TMU) 222, 223, 225–7, 345
Time study 3, 6, 14, **31–46**, 124, 164, 165, 345
  equipment 7
  filing 59
  observation 32–9
  of process-controlled work 164–5
  process 141
  programmes 48–9, 58–9
  summarizing a 39–43
Time values (see standard times)
  register of 182–3
Timing methods—
  by wrist watch or wall clock 2, 318–21
  cumulative 32, 37
  fly-back 32, 37
Total process work cycle 137–9
Total standard cost 266
Two-dimensional flow charts 208
Two-handed process charts 203

Uncontrolled work 263, 268, 345
Universal Office Controls (UOC) 238
Universal Maintenance Standards (UMS) 239
Universal Standards Data (USD) 232
Unmeasured work (see uncontrolled work)
Unoccupied time (UT) **125–35**, 150, 165, 345
  on a single machine or process 125–8, 148–9
  on machine controlled work 137, 138
  on multiple-machine work 151–7, 169–72
  on process-controlled work 153, 154, 156
  per cycle 148, 150–4, 156
  per number of pieces 130–1
  percentage 149, 153, 154, 156
  plus machine interference 159, 161, 169–71
  reduction of, on a "flow-line" team 131, 133, 136
  reduction of, on a team of two 131–2

Unoccupied Time—*contd.*
  reduction of, on multiple machine work 149–51
  reduction of, on process controlled work 125–8, 149–51
  total 133
Utilization index, machine 149, 153, 154, 156, 159

Value analysis 336–7
Variable elements 78, 79, 176, 345
Variable overheads 272, 273
Variety reduction 3, 336

Waiting time 90, 91, 165, 242, 245, 263, 345
  payment for 2, 42, 244, 245
Wink counter 194
Work cycle time, operator 148, 151–4
Work elements (see elements)
Work Factor 219–23, 345
Work measurement 1–4, 345
  broad 95–124
  by analytical observation 96, **117–23**, 340
  by examining work records 95, **98–9**, 318–21
  by using predetermined motion-time systems 2, 3, 6, 96, 122, **219–39**, 294, 344
  by using rated activity sampling 116–117, 344
  by using time study 3, 6, 14, **31–46**, 124, 164, 165, 345
  by wrist watch or wall clock 95, **96–7**, 124
  directly, without a time piece 95, **98**, 123
  equipment 6, 7, 31, 32
  estimating time necessary for 184–5
  fine, for synthetics 315–18
  of long cycle work 326
  of machine operated work 323–4
  of non-recurrent work 318, 325–6
  of process-controlled work 324–5
  of repetitive work 322–6
  of team work **125–36**, 324
  use of films for 4, 219
Work specifications **173–6**, 311, 345
Work study 1, 13
  adaptations of procedure 322–32
  announcing 6
  broad 2, 3
  department 1–13
  equipment 6, 7, 31, 32
  methods of using 4
  procedure in outline 294–8
  procedure in detail 298, 314
  referencing procedure 315–22
  staff 7–9
Work study observation and record sheet 32, **34**, 36, 40, **41**, 97, 315, 319
Work study observation sheet 32, **33,** 36, 96, 117, 140, 141, 143, 197, 321
Work study top sheet 42, 43, **44, 51**, 54, 97, 117, 123, 140, 142, 315, 316, 318, 319, 321
Work unit 29, 30, 91, 345
Work values (see standard times)